SOLUTIONS MANUAL

for the

ENGINEER-IN-TRAINING REVIEW MANUAL

with a

Sample Examination

Michael R. Lindeburg, P.E.

PROFESSIONAL PUBLICATIONS, INC.
Belmont, CA 94002

In the **ENGINEERING REVIEW MANUAL SERIES**

Engineer-In-Training Review Manual
  Engineering Fundamentals Quick Reference Cards
  Mini-Exams for the E-I-T Exam
  1001 Solved Engineering Fundamentals Problems
  E-I-T Review: A Study Guide
Civil Engineering Reference Manual
  Civil Engineering Quick Reference Cards
  Civil Engineering Sample Examination
  Civil Engineering Review Course on Cassettes
  Seismic Design for the Civil P.E. Exam
  Timber Design for the Civil P.E. Exam
Structural Engineering Practice Problem Manual
Mechanical Engineering Review Manual
  Mechanical Engineering Quick Reference Cards
  Mechanical Engineering Sample Examination
  101 Solved Mechanical Engineering Problems
  Mechanical Engineering Review Course on Cassettes
  Consolidated Gas Dynamics Tables
Electrical Engineering Review Manual
Chemical Engineering Reference Manual
  Chemical Engineering Practice Exam Set
Land Surveyor Reference Manual
Metallurgical Engineering Practice Problem Manual
Petroleum Engineering Practice Problem Manual
Expanded Interest Tables
Engineering Law, Design Liability, and Professional Ethics
Engineering Unit Conversions

In the **ENGINEERING CAREER ADVANCEMENT SERIES**

How to Become a Professional Engineer
The Expert Witness Handbook—A Guide for Engineers
Getting Started as a Consulting Engineer
Intellectual Property Protection—A Guide for Engineers
E-I-T/P.E. Course Coordinator's Handbook

Distributed by: Professional Publications, Inc.
1250 Fifth Avenue
Department 77
Belmont, CA 94002
(415) 593-9119

SOLUTIONS MANUAL for the
ENGINEER-IN-TRAINING REVIEW MANUAL

Printed in the United States of America

ISBN: 0-932276-49-0

Professional Publications, Inc.
1250 Fifth Avenue, Belmont, CA 94002

Current printing of this edition (last number): 12 11 10 9

21 Solutions for practice problems in the ENGINEER-IN-TRAINING REVIEW MANUAL

The Engineer-In-Training examination is a difficult examination. To aid you with and to substantially speed up your preparation, the Professional Engineering Registration Program has developed this set of solutions. Although this set of solutions was intended originally for instructors in charge of classroom E-I-T review courses, you may use it as part of your self-study program.

A successful self-study program will possess all of the following characteristics:

1. Exposure to all exam subjects
2. Organized study schedule
3. Regular study hours
4. Proper study method

Exposure to all of the examination subjects is necessary because many of the exam questions are multi-disciplinary. A fluids problem may have elements of thermodynamics. A mechanics of materials problem may have elements of statics or materials science. And so on. Therefore, it is not sufficient to study only a few of your favorite subjects.

It is important that you organize your study schedule and then stick to it. You may wish to use the outline in the introduction as a guide. The *ENGINEER-IN-TRAINING REVIEW MANUAL* and its end-of-chapter problems were designed around a course with 14 weekly meetings of 3 hours each. Therefore, it is suggested that you devote at least a week to each of the subjects on the study outline.

It is suggested that you spread your study time throughout the entire week. This is more conducive to retention than all-day Saturday study sessions. Depending on your familiarity with the subjects, you should plan on one to ten hours of study time each week.

Proper use of the solutions dictates that they be referred to only after you have attempted a problem. Retention of important data, formulas, and solution techniques will be severely compromised if you skim the solutions without actually trying the problems.

SOLUTIONS TABLE OF CONTENTS

E-I-T Homework Solutions: MATHEMATICS

1

4" 4" α|α 1.33 2.67 4"

$\alpha = \text{ARCCOS}\ \frac{1.33}{4} = 1.23 \text{ RAD}$

$\phi = 2\alpha = 2.46 \text{ RAD}$

FROM EQN 1.3,

$A = \frac{1}{2}r^2(\phi - \sin\phi)$

{REMEMBER, ϕ IS IN RADIANS}

$A = (.5)(4)^2(2.46 - \sin 2.46) = 14.64 \text{ IN}^2$

2

4' 3'

AREA OF BASE:

$A_b = \frac{\pi}{4}d^2 = \frac{\pi}{4}(3)^2 = 7.07 \text{ FT}^2$

FROM EQN 1.27,

$A_s = \pi r\sqrt{r^2 + h^2}$

$= \pi(1.5)\sqrt{(1.5)^2 + (4)^2} = 20.13$

TOTAL AREA: $7.07 + 20.13 = 27.20 \text{ FT}^2$

3

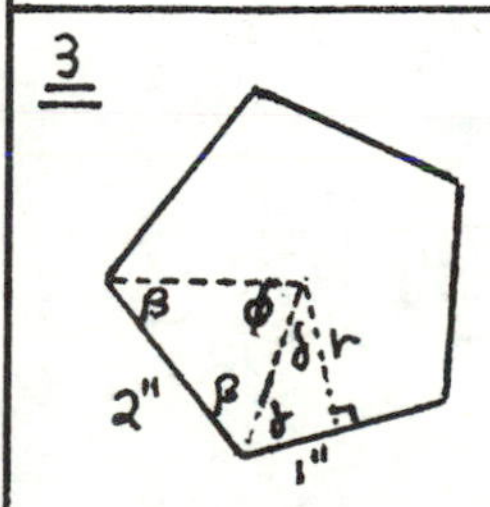

FROM EQN 1.19,

$\phi = \frac{2\pi}{N} = \frac{360°}{5} = 72°$

$\beta = \frac{1}{2}(180 - 72) = 54°$

$\gamma = \beta = 54°$

$\delta = 180 - 90 - 54° = 36°$

$r = \frac{1}{\tan 36°} = 1.376$

$d = 2r = 2(1.376) = 2.753"$

4

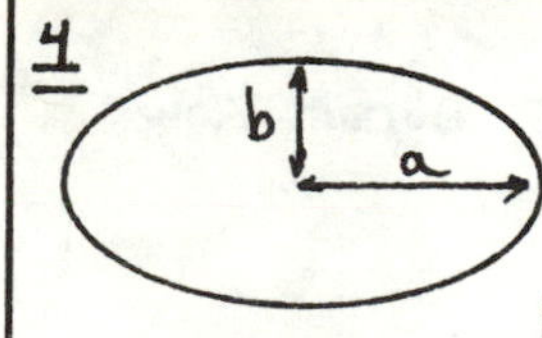

$\frac{a}{b} = 2 \quad \text{OR} \quad a = 2b$

FROM EQN 1.11,

$p = 2\pi\sqrt{\frac{1}{2}(a^2 + b^2)}$

$18 = 2\pi\sqrt{\frac{1}{2}\left[(2b)^2 + b^2\right]}$

OR $b = 1.812$

$a = 2b = 3.624$

FROM EQN 1.10,

$A = \pi ab = \pi(3.624)(1.812) = 20.63 \text{ IN}^2$

5

a) ALL ARE SIGNIFICANT: 5

b) IF THE DECIMAL ZEROS AREN'T SIGNIFICANT, THEY SHOULDN'T BE THERE: 7

c) NOT THE LEADING ZEROS: 4

d) JUST THE 7.93: 3

6

a) $22.52 \rightarrow 22.5$ WITH 3 SIGNIFICANT DIGITS

b) $.11346 \rightarrow .1$ WITH 1 S.D.

c) $3.4913 \rightarrow 3$ WITH 1 S.D.

d) $-.00229 \rightarrow -.002$ WITH 1 S.D.

7

SOLVED BY COMPLETING THE SQUARE UNLESS NOTED OTHERWISE:

a) $f(x) = x^2 + 6x + 8$

$x^2 + 6x = -8$

$(x+3)^2 = -8 + 9$

$x + 3 = \pm\sqrt{1} \qquad x = -2, -4$

$f(x) = (x+2)(x+4)$

b) $f(x) = 3x^3 - 3x^2 - 18x$

$= 3x\left[x^2 - x - 6\right]$

SO, $x = 0$ IS A ROOT. NOW, WORK WITH THE QUADRATIC PART IN BRACKETS.

$x^2 - x - 6 = 0$

$x^2 - x = 6$

$(x - \frac{1}{2})^2 = 6 + \frac{1}{4}$

$x - \frac{1}{2} = \pm\sqrt{6.25}$

$x = 3, -2$

SO $f(x) = 3x(x+2)(x-3)$

c) $f(x) = x^4 + 7x^2 + 12$

$x^4 + 7x^2 = -12$

$(x^2 + 3.5)^2 = -12 + 12.25$

$(x^2 + 3.5) = \pm\sqrt{.25}$

$x^2 = -4, -3$

$f(x) = (x^2+3)(x^2+4)$

d) $f(x) = x^2 - 10x + 16$

$x^2 - 10x = -16$

$(x-5)^2 = -16 + 25 = 9$

$x - 5 = \sqrt{9} = \pm 3$

$x = 5 \pm 3 = 8, 2$

$f(x) = (x-8)(x-2)$

8 THESE ARE OF THE FORM

$$(a+b)(a-b) = a^2 - b^2$$

AND ARE SOLVED BY INSPECTION

a) $(x+3)(x-3)$

b) $3(x^2-4) = 3(x+2)(x-2)$

c) $(1-x^8) = (1+x^4)(1-x^4) = (1+x^4)(1+x^2)(1-x^2)$

$= (1+x^4)(1+x^2)(1+x)(1-x)$

d) $(x^2+y^2)(x^2-y^2) = (x^2+y^2)(x+y)(x-y)$

9 THESE ARE PERFECT SQUARES SOLVED BY INSPECTION

a) $(x+4)^2$

b) $(x-2)^2$

c) $(2y+1)^2$

d) $(6y+5x)^2$

10 MOST OF THESE ARE OF THE FORM

$$a^3 \pm b^3 = (a \pm b)(a^2 \mp ab + b^2)$$

a) $x^3+8 = x^3+(2)^3 = (x+2)(x^2-2x+4)$

b) $(x^6-y^6) = \left[(x^2)^3 - (y^2)^3\right]$

$= (x^2-y^2)(x^4+x^2y^2+y^4)$

$= (x+y)(x-y)(x^2+xy+y^2)(x^2-xy+y^2)$

c) $(x-2)^3 - 8y^3 = (x-2)^3 - (2y)^3$

$= (x-2-2y)\left[(x-2)^2 + (x-2)(2y) + (2y)^2\right]$

d) $6x^2 - 4ax - 9bx + 6ab$

$= 6x^2 - (4a+9b)x + 6ab$

$= (2x-3b)(3x-2a)$

e) $(64+y^3) = (4)^3 + y^3 = (4+y)(4^2-4y+y^2)$

f) $(z^5+32) = z^5 + 2^5$

$= (z+2)(z^4 - 2z^3 + 4z^2 - 8z + 16)$

11 a) $4x^2 - 12x = -7$

$x^2 - 3x = -\frac{7}{4}$

$(x-1.5)^2 = -\frac{7}{4} + (1.5)^2$

$x - 1.5 = \pm\sqrt{.5}$

$x = 2.21,\ .79$

b) $2x^2 - 400 = 0$

$x^2 = 200$

$x = \pm\sqrt{200} = \pm 10\sqrt{2}$

c) $3x^2 = 9 - 36 = -27$

$x^2 = \sqrt{-9}$

$x = \pm 3i$

d) USE THE QUADRATIC ON THIS ONE

$6x^2 - 7x - 5 = 0$

$$x_1, x_2 = \frac{-b \pm \sqrt{b^2-4ac}}{2a}$$

$$= \frac{7 \pm \sqrt{49-(4)(6)(-5)}}{12} = \frac{7 \pm 13}{12}$$

$= 5/3,\ -1/2$

12 a) $4x + 2y = 5$

$5x - 3y = -2$

MULTIPLY TOP ROW BY 5, BOTTOM ROW BY -4 AND ADD

$20x + 10y = 25$

$-20x + 12y = 8$

$0 + 22y = 33$ ADD

SO $y^* = \frac{33}{22} = \frac{3}{2}$

FROM THE TOP EQUATION,

$4x = 5 - 2(3/2)$ OR $x^* = 1/2$

b) $3x - y = 6$

$9x - y = 12$

$-6x + 0 = -6$ SUBTRACT

$x^* = 1$

FROM TOP EQUATION

$y^* = 3(1) - 6 = -3$

c) USE CRAMER'S RULE (PAGE 1-5) TO SOLVE 3x3's

$$\begin{vmatrix} 2 & -2 & 3 \\ 1 & -3 & -2 \\ 1 & 1 & 1 \end{vmatrix} = 16 \qquad \begin{vmatrix} 1 & -2 & 3 \\ -9 & -3 & -2 \\ 6 & 1 & 1 \end{vmatrix} = 32$$

$$\begin{vmatrix} 2 & 1 & 3 \\ 1 & -9 & -2 \\ 1 & 6 & 1 \end{vmatrix} = 48 \qquad \begin{vmatrix} 2 & -2 & 1 \\ 1 & -3 & -9 \\ 1 & 1 & 6 \end{vmatrix} = 16$$

$$x^* = \frac{32}{16} = 2 \qquad y^* = \frac{48}{16} = 3 \qquad z^* = \frac{16}{16} = 1$$

d)

$$\begin{vmatrix} 1 & 1 & 0 \\ 1 & 0 & 1 \\ 3 & -1 & 2 \end{vmatrix} = 2 \qquad \begin{vmatrix} -4 & 1 & 0 \\ 1 & 0 & 1 \\ 4 & -1 & 2 \end{vmatrix} = -2$$

$$\begin{vmatrix} 1 & -4 & 0 \\ 1 & 1 & 1 \\ 3 & 4 & 2 \end{vmatrix} = -6 \qquad \begin{vmatrix} 1 & 1 & -4 \\ 1 & 0 & 1 \\ 3 & -1 & 4 \end{vmatrix} = 4$$

$$x^* = \frac{-2}{2} = -1 \qquad y^* = \frac{-6}{2} = -3 \qquad z^* = \frac{4}{2} = 2$$

13 a) USE SUBSTITUTION

$$2x^2 - (x-1)^2 - 14 = 0$$
$$2x^2 - x^2 + 2x - 1 - 14 = 0$$
$$x^2 + 2x = 15$$
$$(x+1)^2 = 15 + 1$$
$$(x+1) = \pm 4 \qquad x = 3, -5$$

AND $y = x - 1$, SO $y = 2, -6$

$$(x^*, y^*) = (3,2), (-5,-6)$$

b)

$$y - 3x + 4 = 0$$
$$y + \frac{x^2}{y} = \frac{24}{y}$$

REARRANGE THE TOP EQUATION AND MULTIPLY BOTTOM EQUATION BY Y

$$y = 3x - 4$$
$$y^2 + x^2 = 24$$

NOW SUBSTITUTE

13b CONTINUED

$$(3x-4)^2 + x^2 = 24$$
$$9x^2 - 24 + 16 + x^2 = 24$$
$$10x^2 - 24x = 8$$
$$x^2 - 2.4x = .8$$
$$(x - 1.2)^2 = .8 + 1.44$$
$$x - 1.2 = \pm\sqrt{2.24}$$
$$x = 2.697, -.297$$

NOW, $y = 3x - 4$, SO $y = 3(2.697) - 4 = 4.091$

AND $y = 3(-.297) - 4 = -4.891$

SO $(x^*, y^*) = (2.697, 4.091), (-.297, -4.891)$

c)

$$x^2 + y^2 = 25$$
$$x = 10 - 2y$$
$$(10-2y)^2 + y^2 = 25$$
$$100 - 40y + 4y^2 + y^2 = 25$$
$$5y^2 - 40y + 100 = 25$$
$$y^2 - 8y = -15$$
$$(y-4)^2 = -15 + 16$$
$$y - 4 = \pm 1 \qquad y = 5, 3$$

$x = 10 - 2y$, SO $x = 10 - 2(5) = 0$

$x = 10 - 2(3) = 4$

SO $(x^*, y^*) = (0,5), (4,3)$

d) MULTIPLY THE TOP EQUATION BY 3 AND THE BOTTOM EQUATION BY -2

$$6x^2 - 9y^2 = 18$$
$$-6x^2 - 4y^2 = -70$$

$-13y^2 = -52$ ADD

$$y^2 = \frac{52}{13} = 4$$
$$y = \pm 2$$

FROM THE TOP EQUATION

$$x = \sqrt{3 + \tfrac{3}{2}y^2} = \pm 3$$

so $(x^*, y^*) = (\pm 3, \pm 2)$

14 $\log[38.5^x] = \log[6.5^{x-2}]$

$x(\log 38.5) = (x-2)(\log 6.5)$

$x(1.585) = (x-2)(.812)$

$(1.585)x = .812x - 1.624$

$.773x = -1.624$

$x = -2.10$

15

LOG	CHARACTERISTIC	MANTISSA
1.60853	1	.60853
0.96755	0	.96755
−2.14874	−3	.85126
3.60969	3	.60969

16 $\log(1.4) = 1.32[\log(.0613/x)]$

$\log(1.4) = 1.32[\log(.0613) - \log(x)]$

$.1461 = 1.32[-1.2125 - \log(x)]$

$.1107 = [-1.2125 - \log(x)]$

$\log x = -1.2125 - .1107 = -1.3232$

$x = .04751$

17 FOUND BY INSPECTION

a) $\dfrac{1}{2i-1}$

b) $\dfrac{i+3}{(i+1)^2-1} = \dfrac{i+3}{i^2+2i} = \dfrac{i+3}{i(i+2)}$

c) $\dfrac{2i-1}{2i+3}$

d) $3i-1$

18 a) ALL TERMS ARE LESS THAN THE P-SERIES (PAGE 1-8) WITH $p=2$. SINCE P-SERIES CONVERGES, THIS SERIES CONVERGES.

b) SAME AS A HARMONIC SERIES (PAGE 1-8) WITH $a=d=2$. SINCE THE HARMONIC SERIES DIVERGES, THIS SERIES DIVERGES

c) USE THE RATIO TEST (PAGE 1-8)

18 C CONTINUED

$$\lim_{n\to\infty} \frac{b_{n+1}}{b_n} = \frac{\frac{2^{n+1}}{(n+1)!}}{\frac{2^n}{n!}} = \lim_{n\to\infty}\left(\frac{2}{n+1}\right) = 0$$

SINCE LIM → 0, THIS SERIES CONVERGES

d) THIS SERIES IS SMALLER THAN AN ALL-POSITIVE SERIES. AN ALL-POSITIVE SERIES IS SMALLER THAN A P-SERIES WITH $p=2$, SO THE ALTERNATING SERIES CONVERGES

e) THE SERIES DIVERGES. THE TERMS ALTERNATE IN SIGN AND APPROACH ±1.

19

a) $\dfrac{x+2}{x^2-7x+12} = \dfrac{x+2}{(x-3)(x-4)}$ CASE 1

FROM PAGE 1-7, THIS CAN BE EXPRESSED IN THE FORM

$$\frac{A_1}{(x-3)} + \frac{A_2}{(x-4)}$$

$$\frac{A_1(x-4)}{(x-3)(x-4)} + \frac{A_2(x-3)}{(x-4)(x-3)}$$

$$\frac{A_1x - 4A_1 + A_2x - 3A_2}{(x-3)(x-4)}$$

SO $(A_1+A_2)x - (4A_1+3A_2) = x+2$

$$\left.\begin{aligned} A_1 + A_2 &= 1 \\ -4A_1 - 3A_2 &= 2 \end{aligned}\right\} \begin{aligned} A_1 &= -5 \\ A_2 &= 6 \end{aligned}$$

SO $\dfrac{-5}{x-3} + \dfrac{6}{x-4}$

b) $\dfrac{5x+4}{x^2+2x} = \dfrac{5x+4}{(x)(x+2)}$ CASE 1

$$\frac{A_1}{x} + \frac{A_2}{x+2} = \frac{A_1(x+2)}{x(x+2)} + \frac{A_2 x}{(x+2)x}$$

SO $A_1(x+2) + A_2(x) = 5x+4$

$2A_1 = 4$ OR $A_1 = 2$

$A_1 + A_2 = 5$ SO $A_2 = 3$

$$\frac{2}{x} + \frac{3}{x+2}$$

c) $$\frac{3x^2-8x+9}{(x-2)(x-2)(x-2)} = \frac{A_1}{(x-2)} + \frac{A_2}{(x-2)(x-2)} + \frac{A_3}{(x-2)(x-2)(x-2)}$$

CASE 2

$$\frac{A_1(x-2)^2 + A_2(x-2) + A_3}{(x-2)^3}$$

$$= \frac{A_1x^2 - 4A_1x + 4A_1 + A_2x - 2A_2 + A_3}{(x-2)^3}$$

$$= \frac{A_1x^2 + (A_2 - 4A_1)x + 4A_1 - 2A_2 + A_3}{(x-2)^3}$$

so $A_1 = 3$

$A_2 - 4A_1 = -8$ or $A_2 = 4$

$4A_1 - 2A_2 + A_3 = 9$ or $A_3 = 5$

$$\frac{3}{(x-2)} + \frac{4}{(x-2)^2} + \frac{5}{(x-2)^3}$$

d) $$\frac{3x}{x^3-1} = \frac{3x}{(x-1)(x^2+x+1)}$$ CASES 1, 3

$$= \frac{A_1}{x-1} + \frac{A_2x + B_2}{x^2+x+1}$$

$$= \frac{A_1(x^2+x+1) + (A_2x + B_2)(x-1)}{(x-1)(x^2+x+1)}$$

$$\frac{A_1x^2 + A_1x + A_1 + A_2x^2 - A_2x + B_2x - B_2}{(x-1)(x^2+x+1)}$$

so $x^2(A_1 + A_2) + x(A_1 - A_2 + B_2) + (A_1 - B_2) = 3x$

$$\left.\begin{aligned} A_1 + A_2 &= 0 \\ A_1 - A_2 + B_2 &= 3 \\ A_1 - B_2 &= 0 \end{aligned}\right\} \begin{aligned} A_1 &= 1 \\ A_2 &= -1 \\ B_2 &= 1 \end{aligned}$$

$$\frac{1}{x-1} + \frac{1-x}{x^2+x+1}$$

20

a) $\begin{bmatrix} 4 & 12 \\ 3 & 7 \end{bmatrix}$ by inspection

b) $\begin{bmatrix} 6 & 6 & 6 \\ 5 & 3 & 1 \\ 9 & 8 & 1 \end{bmatrix}$ by inspection

c) $$\begin{bmatrix} (2)(3)+(1)(1) & (2)(2)+(1)(1) \\ (0)(3)+(0)(1) & (0)(2)+(0)(1) \end{bmatrix} = \begin{bmatrix} 7 & 5 \\ 0 & 0 \end{bmatrix}$$

d) $$\begin{bmatrix} (1)(2)+(9)(4) & (1)(1)+(9)(-1) & (1)(3)+(9)(-7) \\ (7)(2)+(2)(4) & (7)(1)+(2)(-1) & (7)(3)+(2)(-7) \end{bmatrix}$$

$$= \begin{bmatrix} 38 & -8 & -60 \\ 22 & 5 & 7 \end{bmatrix}$$

21

a) $(3)(-4) - (2)(1) = -14$

b) $(4)(1) - (1)(-6) = 10$

c) EXPAND BY 2ND ROW

$$-(-1)\begin{vmatrix} 1 & 2 \\ 4 & -2 \end{vmatrix} = (1)(-2) - (4)(2) = -10$$

d) EXPAND BY FIRST COLUMN

$$2\begin{vmatrix} 1 & 2 \\ -2 & 1 \end{vmatrix} + 3\begin{vmatrix} -1 & 3 \\ 1 & 2 \end{vmatrix}$$

$$= 2(5) + 3(-5) = -5$$

22 SEE PAGE 1-12,

a) USE THE 2x2 PROCEDURE

DETERMINANT $= (6)(4) - (2)(1) = 22$

$$A^{-1} = \frac{1}{22}\begin{bmatrix} 4 & -1 \\ -2 & 6 \end{bmatrix}$$

b) USE THE 3x3 PROCEDURE

EXPAND BY 3RD COLUMN TO GET DETERMINANT

$$(1)\begin{vmatrix} 1 & 3 \\ -1 & 4 \end{vmatrix} - (1)\begin{vmatrix} 2 & 1 \\ -1 & 4 \end{vmatrix} = (7) - (9) = -2$$

THE COFACTOR MATRIX WITH APPROPRIATE SIGNS IS

$$\begin{bmatrix} -4 & -1 & 7 \\ 4 & 1 & -9 \\ -2 & -1 & 5 \end{bmatrix}$$

THE TRANSPOSED COFACTOR MATRIX IS

$$\begin{bmatrix} -4 & 4 & -2 \\ -1 & 1 & -1 \\ 7 & -9 & 5 \end{bmatrix}$$

22 b CONTINUED

DIVIDE ALL ENTRIES BY -2 (THE DETERMINANT) TO GET THE INVERSE

$$\begin{bmatrix} 2 & -2 & 1 \\ 1/2 & -1/2 & 1/2 \\ -3.5 & 4.5 & -2.5 \end{bmatrix}$$

23

a) $\theta = \text{ARCTAN}\ \frac{9.6}{17.2} = 29.2°$

b) $\theta = \text{ARCSIN}\ \frac{3.1}{5.4} = 35.0°$

c) $\theta = \text{ARCCOS}\ \frac{2.9}{14.1} = 78.1°$

24

a) $\sec X - (\sec X)(\sin^2 X)$

$\sec X [1 - \sin^2 X] = \frac{1}{\cos X}[\cos^2 X] = \cos X$

b) $\sin^2 X (1 + \cot^2 X) = \sin^2 X (\csc^2 X)$

$= \frac{\sin^2 X}{\sin^2 X} = 1$

c) $\tan^2 X \cos^2 X + \cot^2 X \sin^2 X$

$\frac{\sin^2 X \cos^2 X}{\cos^2 X} + \frac{\cos^2 X \sin^2 X}{\sin^2 X}$

$= \sin^2 X + \cos^2 X = 1$

25

$C = 180° - 32° - 70° = 78°$

a) USE THE LAW OF SINES

$\frac{27}{\sin 78°} = \frac{b}{\sin 70°} = \frac{a}{\sin 32°}$

$a = 14.6 \quad b = 25.9$

b) $C = 180 - 25 - 40 = 115°$

USE THE LAW OF SINES

$\frac{63}{\sin 115°} = \frac{b}{\sin 40°} = \frac{a}{\sin 25°}$

$a = 29.4 \quad b = 44.7$

26

FIND THE SLOPE

$M = \frac{\Delta Y}{\Delta X} = \frac{9.5 - 3.4}{8.3 - 1.7} = .924$

SUBSTITUTE ANY POINT INTO THE SLOPE FORM TO FIND b

$3.4 = .924(1.7) + b \rightarrow b = 1.83$

$y = .924x + 1.83$

b) $M = \frac{-7.2 - (-3.8)}{5.1 - (-6)} = -.3063$

$-3.8 = -.3063(-6) + b \rightarrow b = -5.638$

$y = -.306x - 5.64$

27

FROM PAGE 1-14,

FOR $y = .924x + 1.83$

a) $.924x - y + 1.83 = 0$

b) $(y - 3.4) = .924(x - 1.7)$

c) DIVIDE ALL TERMS BY 1.83

$\frac{y}{1.83} - \frac{x}{1.98} = 1$

FOR $y = -.306x - 5.64$

a) $-.306x - y - 5.64 = 0$ OR $.306x + y + 5.64 = 0$

b) $(y + 7.2) = -.306(x - 5.1)$

c) DIVIDE THROUGH BY -5.64

$\frac{-x}{18.4} - \frac{y}{5.64} = 1$

28

$\theta = \text{ARCTAN}(M) = \text{ARCTAN}(.4) = 21.8°$

29

FROM EQN 1.127,

$d = \sqrt{(1-(-4))^2 + (-3-5)^2 + (9-(-7))^2} = 18.57$

30

THE LINE PASSING THROUGH (3, -2) AND (-6, 5) HAS THE SLOPE

$M = \frac{5 - (-2)}{-6 - 3} = -.777$

THE PERPENDICULAR LINE HAS A SLOPE OF

$\frac{-1}{M} = 1.286$

$1 = 1.286(3) + b \longrightarrow b = -2.858$

$y = 1.286x - 2.858$

31

$\phi = \text{ARCSIN}\ 3/5 = 36.87$

$\tan \phi = -.75$

$y = mx + b$

$0 = (-.75)x + b \rightarrow b = 3.75$

$y = -.75x + 3.75$

OR

$y = .75x - 3.75$

32

$(-6,8)$

$M = -\frac{8}{6} = -1.33$

THE LINE SLOPE IS

$\frac{-1}{-1.33} = .75$

$y = Mx + b$

$8 = (.75)(-6) + b \rightarrow b = 12.5$

$y = .75x + 12.5$

33

(a) $(2)(5) + (3)(-2) = 10 - 6 = 4$

(b) $(2)(5) + (3)(-2) = 10 - 6 = 4$

(c) $(2)(8) + (-3)(2) + (6)(-3) = -8$

34

$(2,3)$, $(5,-2)$

$\phi = \text{ARCTAN}\left(\frac{3}{2}\right) + \text{ARCTAN}\left(\frac{2}{5}\right) = 78.1°$

35

a) $\begin{vmatrix} i & 1 & 9 \\ J & 4 & -3 \\ K & 0 & 0 \end{vmatrix}$ EXPAND BY 3RD ROW

$K[(1)(-3) - (4)(9)] = -39K$

b) $\begin{vmatrix} i & 6 & 1 \\ J & 2 & 0 \\ K & 3 & 1 \end{vmatrix}$ EXPAND BY 3RD COLUMN

$[J3 - 2K] + [2i - 6J] = 2i - 3J - 2K$

c) $\begin{vmatrix} 1 & 7 & 3 \\ J & -3 & 4 \\ K & 0 & 0 \end{vmatrix}$ EXPAND BY 3RD ROW

$K[28 - (-9)] = 37K$

36

a) $L = 9 \quad M = 7$

$d = \sqrt{(9)^2 + (7)^2} = 11.40$

$\cos\alpha = \frac{9}{11.40} = .789$

$\cos\beta = \frac{7}{11.40} = .614$

b) $L = 3 \quad M = 2 \quad N = 6$

$d = \sqrt{(3)^2 + (2)^2 + (-6)^2} = 7$

$\cos\alpha = 3/7 = .429$

$\cos\beta = 2/7 = .286$

$\cos\gamma = -6/7 = -.857$

c) $L = -2 \quad M = 1 \quad N = -3$

$d = \sqrt{(-2)^2 + (1)^2 + (-3)^2} = 3.74$

$\cos\alpha = \frac{-2}{3.74} = -.535$

$\cos\beta = \frac{1}{3.74} = .267$

$\cos\gamma = \frac{-3}{3.74} = -.802$

37

$N = i + 2J - 3K$

$D = -[(1)(3) + (2)(1) + (-3)(2)] = 1$

SO THE PLANE IS

$(x-3) + 2(y-1) - 3(z-2) = 0$

OR $x + 2y - 3z + 1 = 0$

38

LET $V_1 = (2-0)i + (4-0)J + (1-2)K = 2i + 4J - K$

$V_2 = (-2-0)i + (3-0)J + (3-2)K = -2i + 3J + K$

$N = \begin{vmatrix} i & 2 & -2 \\ J & 4 & 3 \\ K & -1 & 1 \end{vmatrix} = i(4+3) + J(2-2) + K(6+8) = 7i + 14K$

THEN $D = -(Ax_0 + By_0 + Cz_0) = -[(7)(0) + (0)(0) + (14)(2)] = -28$

SO THE PLANE IS

$7x + 0y + 14z - 28 = 0$

OR $x + 2z - 4 = 0$

39

$$\cos\phi = \frac{|N_1 \cdot N_2|}{|N_1| \cdot |N_2|} = \frac{|(2)(6) + (-1)(-2) + (-2)(3)|}{\sqrt{(2)^2 + (-1)^2 + (-2)^2}\sqrt{(6)^2 + (-2)^2 + (3)^2}} = .381$$

40

$$d = \frac{(2)(2) + (2)(2) + (-1)(-4) - 6}{\sqrt{(2)^2 + (2)^2 + (-1)^2}} = 2$$

41 REFER TO PAGE 1-17

a)

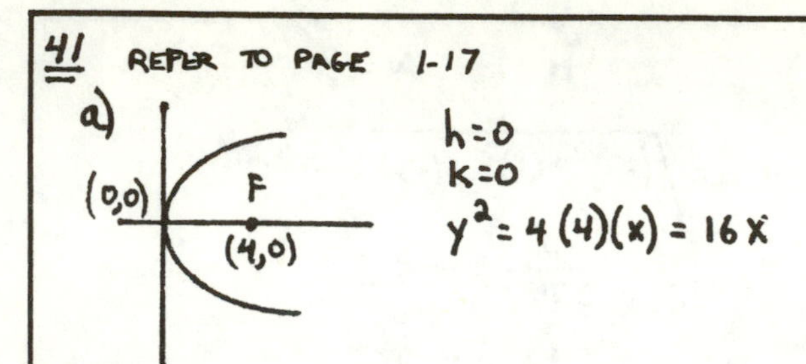

$h = 0$

$k = 0$

$y^2 = 4(4)(x) = 16x$

b)

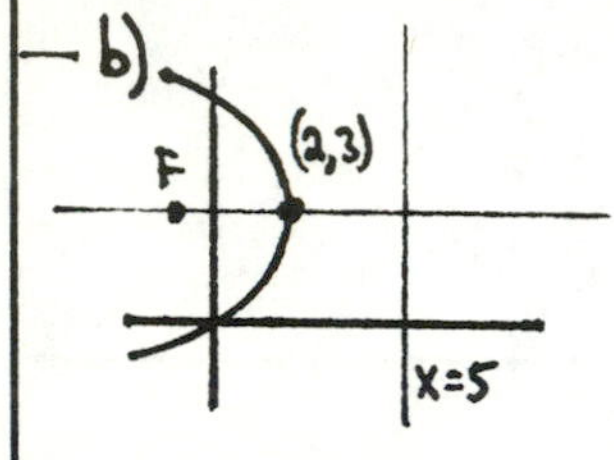

$h = 2$

$k = 3$

DIRECTRIX IS 3 FROM THE VERTEX

SO FOCUS IS AT $(-1,3)$

$(p+2) = -1$ so $p = -3$

$(y-3)^2 = 4(-3)(x-2)$

$(y-3)^2 = -12(x-2)$

c)

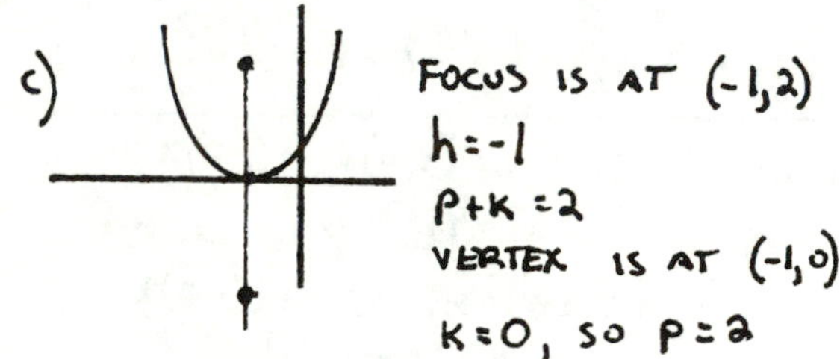

FOCUS IS AT $(-1,2)$

$h = -1$

$p + k = 2$

VERTEX IS AT $(-1,0)$

$k = 0$, so $p = 2$

$(x+1)^2 = 4(2)(y-0)$

$(x+1)^2 = 8y$

42 $9x^2 + 16y^2 + 54x - 64y = -1$

$9x^2 + 54x + 16y^2 - 64y = -1$

$9(x^2 + 6x) + 16(y^2 - 4y) = -1$

COMPLETING THE SQUARE,

$9(x+3)^2 + 16(y-2)^2 = -1 + 81 + 64 = 144$

$$\frac{(x+3)^2}{16} + \frac{(y-2)^2}{9} = 1$$

ELLIPSE

43 a)

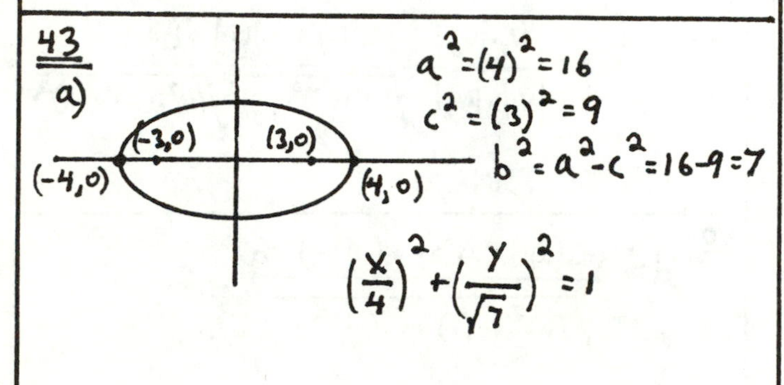

$a^2 = (4)^2 = 16$

$c^2 = (3)^2 = 9$

$b^2 = a^2 - c^2 = 16 - 9 = 7$

$$\left(\frac{x}{4}\right)^2 + \left(\frac{y}{\sqrt{7}}\right)^2 = 1$$

b) (0,4) (0,-4)

$b = \pm 4$

$\epsilon = \frac{1}{a}\sqrt{a^2 - b^2}$

$\sqrt{\frac{33}{49}} = \frac{1}{a}\sqrt{a^2 - 16}$

$\frac{33}{49} = \frac{1}{a^2}(a^2 - 16) \longrightarrow a = 7$

$$\left(\frac{x}{7}\right)^2 + \left(\frac{y}{4}\right)^2 = 1$$

c)

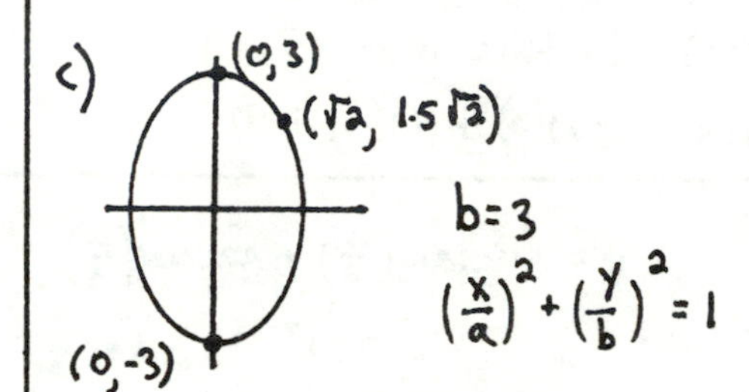

$b = 3$

$\left(\frac{x}{a}\right)^2 + \left(\frac{y}{b}\right)^2 = 1$

SUBSTITUTING FOR THE POINT GIVEN

$$\left(\frac{\sqrt{2}}{a}\right)^2 + \left(\frac{1.5\sqrt{2}}{3}\right)^2 = 1$$

$\frac{2}{a^2} + .5 = 1$ so $a = 2$

$$\left(\frac{x}{2}\right)^2 + \left(\frac{y}{3}\right)^2 = 1$$

44 THIS IS A PERMUTATION OF 6 ITEMS TAKEN 3 AT A TIME

$$P(n,r) = \frac{n!}{(n-r)!} = \frac{6!}{(6-3)!} = \frac{6 \cdot 5 \cdot 4 \cdot 3 \cdot 2 \cdot 1}{3 \cdot 2 \cdot 1}$$

$= 120$

45 a) THE FIRST SEAT CAN BE FILLED 7 DIFFERENT WAYS, THE SECOND SEAT 6 WAYS, ETC.

$7 \cdot 6 \cdot 5 \cdot 4 \cdot 3 \cdot 2 \cdot 1 = 7! = 5040$ WAYS

b) THIS IS THE SAME AS PART (a) EXCEPT THAT THERE IS NO 'FIRST' SEAT

$\frac{5040}{7} = 720$ WAYS

46 THIS IS A PERMUTATION OF N OBJECTS TAKEN N AT A TIME, AND

$$N = N_1 + N_2 + N_3 + \cdots$$

$$p(N,N) = \frac{N!}{N_1!\,N_2!\,N_3!\cdots}$$

$$p(6,6) = \frac{6!}{4!\,2!} = 15$$

47 THIS IS A COMBINATION SINCE THE ORDER IS NOT RELEVANT

a) $C(N,r) = \frac{N!}{(N-r)!\,r!} = \frac{8!}{(8-6)!\,6!} = 28$

b) IF 2 QUESTIONS ARE REQUIRED, HE MUST WORK 4 MORE OUT OF THE 6 REMAINING

$$C(6,4) = \frac{6!}{(6-4)!\,4!} = 15$$

48 THE PROBABILITY DISTRIBUTION IS

HEADS: .6667
TAILS: .3333

a) .6667

b) $(.6667)^3 = .296$

c) $\frac{4!}{(4-2)!\,2!}(.6667)^2(.3333)^2 = .296$

49 THE INDIVIDUAL SUIT PROBABILITY IS $\frac{13}{52}$

a) WHEN THE FIRST SPADE IS DRAWN, THERE REMAINS ONE FEWER CARD AND ONE FEWER SPADE

$$\left(\frac{13}{52}\right)\left(\frac{12}{51}\right) = .0588$$

(b) EITHER A SPADE OR DIAMOND MAY BE FIRST

$$\left(\frac{13}{52}\right)\left(\frac{13}{51}\right) + \left(\frac{13}{52}\right)\left(\frac{13}{51}\right) = .1275$$

c) $\left(\frac{13}{52}\right)\left(\frac{12}{51}\right) + \left(\frac{13}{52}\right)\left(\frac{12}{51}\right) = .1176$

d) $\frac{13}{52}\left[\left(\frac{13}{51} + \frac{13}{51}\right)\right] + \left[\frac{13}{52} + \frac{13}{52}\right]\left(\frac{13}{51}\right) = .2549$

50 THESE ARE TRUE BINOMIAL DISTRIBUTION PROBLEMS WITH $p = .25$, $q = .75$

a) $p(0) = \frac{7!}{(7-0)!\,0!}(.25)^0(.75)^7 = .1335$

b) $p(1) = \frac{7!}{(7-1)!\,1!}(.25)^1(.75)^6 = .3115$

c) $p(2) = \frac{7!}{(7-2)!\,2!}(.25)^2(.75)^5 = .3115$

51 THIS IS A POISSON PROBLEM

$$p(34) = \frac{e^{-38}\,38^{34}}{34!} = .0549$$

52

$$p(2) = \frac{5!}{(5-2)!\,2!}(.10)^2(.90)^3 = .0729$$

53

$$p(2) = \frac{20!}{(20-2)!\,2!}(.15)^2(.85)^{18} = .2293$$

54

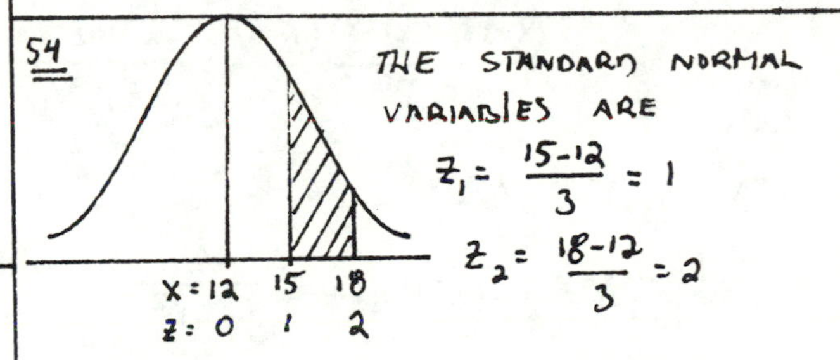

THE STANDARD NORMAL VARIABLES ARE

$$z_1 = \frac{15-12}{3} = 1$$

$$z_2 = \frac{18-12}{3} = 2$$

FROM THE NORMAL TABLE ON PAGE 1-22

$$p\{15 < x < 18\} = .4772 - .3413 = .1359$$

55

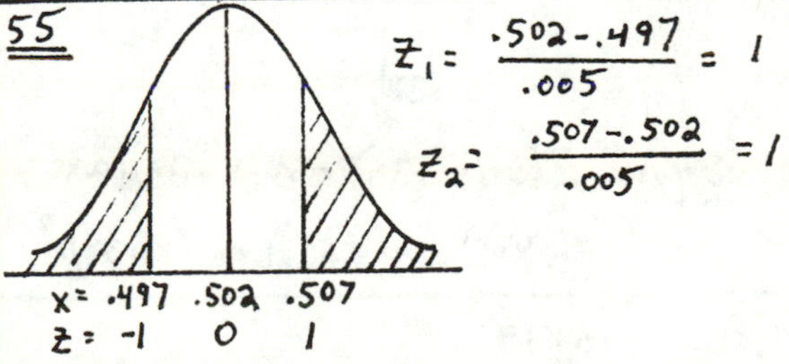

$$z_1 = \frac{.502-.497}{.005} = 1$$

$$z_2 = \frac{.507-.502}{.005} = 1$$

a) $p(\text{DEFECTIVE}) = 2[.5-.3413] = .3174$

b) $p(3,2) = \frac{3!}{(3-2)!\,2!}(.3174)^2(1-.3174)^1 = .2063$

c) $(8)(200)(.3174) = 507.8$

56

$N = 12$

$\sum x = 114$

$\bar{x} = 9.5$

$\sum x^2 = 1428$

$$\sigma^2 = \frac{1428}{12} - (9.5)^2 = 28.75$$

$$\sigma = 5.36$$

57

$N = 3 + 8 + 18 + 12 + 9 = 50$

$\sum x = 3(1.5) + 8(2.5) + 18(3.5) + 12(4.5) + 9(5.5) = 191$

$\sum x^2 = 3(1.5)^2 + 8(2.5)^2 + 18(3.5)^2 + 12(4.5)^2 + 9(5.5)^2 = 792.5$

$$\sigma^2 = \frac{792.5}{50} - \left(\frac{191}{50}\right)^2 = 1.258$$

$\sigma = 1.121$

58

a) $4t + 8$

b) $f(x) = (x+5)(x^2-1)^{-1}$

$f'(x) = (x+5)(-1)(x^2-1)^{-2}(2x) + (x^2-1)^{-1}(1)$

$$= \frac{(x+5)(-2x) + (x^2-1)}{(x^2-1)^2} = \frac{-x^2-10x-1}{(x^2-1)^2}$$

c) $f(x) = 2(x^2-1)^{-1}$

$$f'(x) = -2(x^2-1)^{-2}(2x) = \frac{-4x}{(x^2-1)^2}$$

d) $f(x) = (2-3x^2)^{\frac{1}{2}}$

$$f'(x) = \tfrac{1}{2}(2-3x^2)^{-\frac{1}{2}}(-6x) = \frac{-3x}{\sqrt{2-3x^2}}$$

e) $f(x) = \sin^2(x^2+3x)$

$f'(x) = 2\sin(x^2+3x)\cos(x^2+3x)(2x+3)$

$= (4x+6)\sin(x^2+3x)\cos(x^2+3x)$

59

a) $f(x) = x^3 - 5x - 4$

$f'(x) = 3x^2 - 5$

$f'(x) = 0$ AT ± 1.29

$f(-3) = -16$ MINIMUM

$f(-1.29) = .303$ MAXIMUM

$f(-1) = 0$

b) $f(x) = x^3 + 3x^2 - 9x$

$f'(x) = 3x^2 + 6x - 9 = 0$ AT $x = 1, -3$

$f(-4) = 20$

$f(-3) = 27$

$f(1) = -5$ MINIMUM

$f(4) = 76$ MAXIMUM

c) $f(x) = \dfrac{x+5}{x-3}$ $\quad f'(x) = \dfrac{-8}{(x-3)^2}$

$f'(x)$ IS NEVER ZERO ON THE GIVEN INTERVAL

$f(-5) = 0$ MAXIMUM

$f(2) = -7$ MINIMUM

d) $f(x) = x^3 + 7x^2 - 5x$

$f'(x) = 3x^2 + 14x - 5$

$f'(x) = 0$ AT $x = .333, -5$

$f(-2) = 30$ MAXIMUM

$f(.333) = -.854$ MINIMUM

$f(2) = 26$

60 TREAT ALL NON-X VARIABLES LIKE CONSTANTS

a) $\dfrac{\partial f}{\partial x} = 8x - 3y$

b) $\dfrac{\partial f}{\partial x} = y^2$

c) $\dfrac{\partial f}{\partial x} = 2x$

61

a) USE L'HOPITAL'S RULE TWICE

$$\lim_{x\to 0} \frac{1-\cos x}{x^2} = \lim_{x\to 0} \frac{\sin x}{2x} = \lim_{x\to 0} \frac{\cos x}{2} = \frac{1}{2}$$

b) $$\lim_{x\to 2} \frac{x^2-4}{x-2} = \lim_{x\to 2} \frac{(x+2)(x-2)}{(x-2)} = \lim_{x\to 2}(x+2) = 4$$

c) $$\lim_{x\to 3} \frac{x^3-27}{x-3} = \lim_{x\to 3} \frac{(x-3)(x^2+3x+9)}{x-3}$$

$$= \lim_{x\to 3}(x^2+3x+9) = 27$$

d) DIVIDE ALL TERMS BY X

$$\lim_{x\to\infty} \frac{2x+1}{5x-2} = \lim_{x\to\infty} \frac{2+\frac{1}{x}}{5-\frac{2}{x}} = \frac{2}{5}$$

62 a) USE INTEGRATION BY PARTS.

LET $f = x^2 \quad df = 2x$

$dg = e^x \quad g = e^x$

$$\int x^2 e^x = x^2 e^x - \int e^x 2x$$

USE INTEGRATION BY PARTS AGAIN:

$f = 2x \quad df = 2$

$dg = e^x \quad g = e^x$

$$= x^2 e^x - 2xe^x + \int 2e^x$$

$$= x^2 e^x - 2xe^x + 2e^x = e^x(x^2 - 2x + 2) + C$$

b) $\int (1-x)^{\frac{1}{2}} = -\frac{2}{3}(1-x)^{3/2} + C$ (BY INSPECTION)

c) $\int x(x^2+1)^{-1} = \frac{1}{2}\ln(x^2+1) + C$ (by INSPECTION)

d) $\frac{x^2}{x^2+x-6} = 1 - \frac{x}{x^2+x-6} + \frac{6}{x^2+x-6}$

$$= 1 - \frac{3/5}{x+3} - \frac{2/5}{x-2} - \frac{6/5}{x+3} + \frac{6/5}{x-2}$$

$$\int \frac{x^2}{x^2+x-6}dx = x - \frac{9}{5}\ln(x+3) + \frac{4}{5}\ln(x-2) + C$$

{PROVE BY TAKING DERIVATIVE}

63 a) $\int_1^3 (x^2+4x)\,dx = \left[\frac{1}{3}x^3 + 2x^2\right]_1^3 = 9 + 18 - \frac{1}{3} - 2$

$= 24.667$

b) $\int_{-2}^{2} (x^3+1)\,dx = \left[\frac{1}{4}x^4 + x\right]_{-2}^{2} = 4 + 2 - 4 + 2 = 4$

c) $\int_1^2 (4x^3 - 3x^2)\,dx = \left[x^4 - x^3\right]_1^2 = 16 - 8 - 1 + 1 = 8$

64

y = -x - 1   y = 6x - x²   1   3

$$\text{AREA} = \int_{x=1}^{x=3} 6x - x^2 - (-x-1)$$

$$= \int_1^3 -x^2 + 7x + 1$$

$$= \left[-\frac{1}{3}x^3 + \frac{7}{2}x^2 + x\right]_1^3$$

$$= 21.33$$

65 a) $y' = x^2 - 2x - 4$

$y = \frac{1}{3}x^3 - x^2 - 4x + C$

BUT $y(3) = -6$ SO

$-6 = \frac{1}{3}(3)^3 - (3)^2 - (4)(3) + C$

$C = 6$

$y = \frac{1}{3}x^3 - x^2 - 4x + 6$

b) $y'' + 4y' + 4y = 0$

SOLVE THE CHARACTERISTIC EQUATION (SEE PAGE 1-30):

$x^2 + 4x + 4 = 0 \rightarrow x = -2, -2$

SO $y = a_1 e^{-2x} + a_2 x e^{-2x}$

BUT $y(0) = 1$ SO $a_1 = 1$

AND $y'(0) = 1$

$y' = -2a_1 e^{-2x} + a_2\left[x(-2)e^{-2x} + e^{-2x}\right]$

$1 = (-2)(1)(1) + a_2[0 + 1]$

$1 = -2 + a_2 \rightarrow a_2 = 3$

$y = e^{-2x}(1 + 3x)$

c) $y'' + 3y' + 2y = 0$

SOLVE THE CHARACTERISTIC EQUATION:

$x^2 + 3x + 2 = 0 \rightarrow x = -2, -1$

$y = a_1 e^{-2x} + a_2 e^{-x}$

BUT $y(0) = 1$

$1 = a_1 + a_2$

AND $y'(0) = 0$

$y' = -2a_1 e^{-2x} - a_2 e^{-x}$

$0 = -2a_1 - a_2$

SOLVING SIMULTANEOUSLY,

$a_1 = -1, \ a_2 = 2$

$y = 2e^{-x} - e^{-2x}$

d) USE THE LAPLACE TRANSFORM METHOD

$$y'' + 2y' + 2y = e^{-t}$$

$$\mathcal{L}(y'') + 2\mathcal{L}(y') + 2\mathcal{L}(y) = \mathcal{L}(e^{-t})$$

USING THE INITIAL CONDITIONS,

$$s^2\mathcal{L}(y) - 1 + 2s\mathcal{L}(y) + 2\mathcal{L}(y) = \mathcal{L}(e^{-t})$$

$$\mathcal{L}(y)[s^2+2s+2] - 1 = \frac{1}{s+1}$$

OR

$$\mathcal{L}(y) = \frac{s+2}{(s+1)(s^2+2s+2)}$$

USING PARTIAL FRACTIONS,

$$\frac{s+2}{(s+1)(s^2+2s+2)} = \frac{A_1}{(s+1)} + \frac{A_2 s + B_2}{(s^2+2s+2)}$$

USING THE METHOD OF UNDETERMINED COEFFICIENTS, $A_1 = 1$

$A_2 = -1$

$B_2 = 0$

SO

$$\mathcal{L}(y) = \frac{1}{s+1} - \frac{s}{s^2+2s+2}$$

TAKING THE INVERSE TRANSFORM,

$$y = e^{-t} - e^{-t}\cos(t) + e^{-t}\sin(t)$$

66 LET X = POUNDS OF SALT IN TANK AT TIME t

$X_0 = 60$

X' IS THE RATE AT WHICH THE SALT CONTENT IS CHANGING

2 POUNDS OF SALT ENTER EACH MINUTE

3 GALLONS LEAVE EACH MINUTE, THE SALT LEAVING IS

$$3(\text{CONCENTRATION}) = 3\left(\frac{\text{SALT CONTENT}}{\text{VOLUME}}\right)$$

$$= 3\left(\frac{X}{100-t}\right)$$

SO

$$x'(t) = 2 - 3\left[\frac{x(t)}{100-t}\right]$$

OR

$$x' + \frac{3x}{100-t} = 2$$

THIS IS FIRST ORDER LINEAR (SEE PAGE 1-30)

$$u = \exp\left[3\int\frac{dt}{100-t}\right] = (100-t)^{-3}$$

$$X = (100-t)^3\left[2\int\frac{dt}{(100-t)^3} + K\right]$$

$$= 100 - t + K(100-t)^3$$

BUT $X = 60$ AT $t = 0$ SO

$K = -.00004$

$$X = 100 - t - .00004(100-t)^3$$

$$X(60) = 37.44$$

E-I-T HOMEWORK SOLUTIONS: ENGINEERING ECONOMY

1 THIS IS A 10-YEAR PERIOD

$F = 250(F/P,6\%,10)$

$= 250(1.7908) = 447.70$

2 $P = 2000(P/F,6\%,6)$

$= 2000(.7050) = 1410.00$

3 $P = 2000(P/F,6\%,7)$

$= 2000(.6651) = 1330.20$

4 $A = 50(A/P,6\%,10)$

$= 50(.1359) = 6.80$

5 $F = 20{,}000(F/A,6\%,10)$

$= 20{,}000(13.1808) = 263{,}616.00$

6 $A = 5000(A/F,6\%,7)$

$= 5000(.1191) = 595.50$

7 $P = 400(P/A,6\%,7)$

$= 400(5.5824) = 2232.96$

8 $F = 500(F/P,4\%,10) + 700(F/P,4\%,8) + 900(F/P,4\%,6)$

$= 500(1.4802) + 700(1.3686) + 900(1.2653)$

$= 2836.89$

9 $r = 6\%$ $\quad \phi = \frac{.06}{2} = .03$

$F = 550(1.03)^{16} = 882.59$

10 $75 = 50(F/P,i,5)$

$(F/P,i,5) = \frac{75}{50} = 1.5$

$(F/P,8\%,5) = 1.4693$

$(F/P,10\%,5) = 1.6105$

so $i = 8\% + (10\%-8\%)\frac{1.5-1.4693}{1.6105-1.4693} = 8.43\%$

(F/P,i,5) CAN ALSO BE SOLVED DIRECTLY FOR i, $(1+i)^5 = 1.5-1$, $i = 8.447\%$

11 $r = .04$

THERE ARE $(12)(30) = 360$ MONTHLY COMPOUNDING PERIODS

$\phi = \frac{.04}{12} = .003333$

FROM TABLE 2.1,

$(A/F,.003333,360) = \frac{.003333}{(1.003333)^{360}-1}$

$= .0014409$

$A = (50{,}000)(.0014409) = 72.05$

12 $N = (18)(12) = 216$

FROM TABLE 2.1,

$(F/A,.003333,216) = \frac{(1.003333)^{216}-1}{.003333}$

$= 315.58$

$F = (72.05)(315.58) = 22{,}737.5$

13 ASSUME THE FIRST PAYMENT IS MADE ON THE DAY HE IS BORN AND THE WITHDRAWAL IS ON HIS 18TH BIRTHDAY

$20{,}000 = A(F/P,5\%,18) + A(F/A,5\%,18)$

$= A(2.4066 + 28.1324)$

$A = \frac{20{,}000}{30.539} = 654.90$

14 FIND THE PRESENT WORTH OF BOTH CASH FLOWS AT JANUARY 1974

$50(F/A,6\%,11)(F/P,6\%,4) = X(P/A,6\%,5)$

$50(14.9716)(1.2625) = X(4.2124)$

$X = 224.36$

15 $r = .10$

$\phi = \frac{.10}{12} = .008333$ PER MONTH

MONTH 1

INTEREST $= (2000)(.008333) = 16.67$

PRINCIPLE $= 400.00$

PAYMENT 416.67

MONTH 2

$I = (1600)(.008333) = 13.33$

$P = 400.00$

PAYMENT 413.33

MONTH 3

$I = (1200)(.008333) = 10.00$

$P = 400.00$

PAYMENT 410.00

MONTH 4

$I = (800)(.008333) = 6.67$

$P = 400.00$

PAYMENT 406.67

MONTH 5

$I = (400)(.008333) = 3.33$

$P = 400.00$

403.33

b) 800

c) 6.67

16

$P = -12000 + 2000(P/F, 10\%, 10) - 1000(P/A, 10\%, 10) - 200(P/G, 10\%, 10)$

$= -12000 + 2000(.3855) - 1000(6.1446) - 200(22.8913)$

$= -21951.86$

17 MONEY SAVED PER PIECE IS

$\frac{7(3.90)}{3600} = .007583$

MONEY SAVED PER YEAR

$(40{,}000)(.007583) = 303.32$

$P = 303.32(P/A, 8\%, 3)$

$= 303.32(2.5771) = 781.68$

18

$CC = 100{,}000 + \frac{18{,}000}{.08} = 325{,}000$

19

A: $CC = 120 + \frac{(.07)(400)}{.10} = 400$

B: $CC = 70 + \frac{(.13)(400)}{.10} = 590$

20

$EUAC = (17{,}000 + 5000)(A/P, 6\%, 5) - (14000 + 2500)(A/F, 6\%, 5) + 200$

$= (22000)(.2374) - (16500)(.1774) + 200$

$= 2495.70$

21

$EUAC(\text{Aluminum}) = 6000(A/P, 10\%, 50)$
$= 6000(.1009) = 605.40$

$EUAC(\text{Shingles}) = 3500(A/P, 10\%, 15)$
$= 3500(.1315) = 460.25$

THE SHINGLES ARE BEST

22

$EUAC(A) = 90{,}000(A/P, 12, 20) - 10{,}000(A/F, 12, 20) + 3000 + 2200 + 400$

$= 90{,}000(.1339) - 10{,}000(.0139) + 5600$

$= 17512.00$

$EUAC(B) = 60{,}000(.1339) - 6000(.0139) + 8000$

$= 15950.60$

B IS BEST

23

$14000 = 4000 + 1200(P/A, ?, 12)$

$(P/A, ?, 12) = 8.3333$

FOR 6%, $(P/A) = 8.3838$

FOR 8%, $(P/A) = 7.5361$

$? = 6\% + (8\% - 6\%)\left(\frac{8.3838 - 8.3333}{8.3838 - 7.5361}\right)$

$= 6.1\%$

24 THE ANNUAL RENTAL INCOME IS

$(12)(75) = 900$

$F = (14000 + 1000)(F/P, 10\%, 10) + (150 + 250 - 900)(F/A, 10\%, 10)$

$= 15000(2.5937) - 500(15.9374)$

$= 30936.80$

25

$(40{,}000)(.08) = (10000)(.15) + (30{,}000)X$

$X = .0567$ OR 5.67%

26

$B - C = 500{,}000 - 175{,}000 - 50{,}000$

$= 275{,}000$

$B/C = \frac{500{,}000 - 50{,}000}{175{,}000} = 2.57$

27

$B/C = \frac{1{,}500{,}000 - 300{,}000}{1{,}000{,}000} = 1.2$

$B - C = 1{,}500{,}000 - 300{,}000 - 1{,}000{,}000$

$= 200{,}000$

28 IF KEPT FOR 1 YEAR

$EUAC(1) = 10000(A/P, 8\%, 1) - 6000(A/F, 8\%, 1) + 2300$

$= 10000(1.0800) - 6000(1.000) + 2300$

$= 7100$

IF KEPT FOR 2 YEARS

$EUAC(2) = [10{,}000 + 2300(P/F, 8\%, 1)](A/P, 8\%, 2) + (2500 - 4000)(A/F, 8\%, 2)$

$= [10{,}000 + (2300)(.9259)](.5608) + (-1500)(.4808)$

$= 6081.06$

PROBLEM 28 CONTINUED

IF KEPT FOR 3 YEARS

$$EUAC(3) = [10{,}000 + 2300(P/F,8\%,1) + 2500(P/F,8\%,2)] \times (A/P,8\%,3) + (3300-3200)(A/F,8\%,3)$$
$$= [10{,}000 + (2300)(.9259) + (2500)(.8573)] \times (.3880) + (100)(.3080)$$
$$= 5568.65$$

IF KEPT FOR 4 YEARS

$$EUAC(4) = [10{,}000 + 2300(P/F,8\%,1) + 2500(P/F,8\%,3) + 3300(P/F,8\%,3)](A/P,8\%,4) + (4800-2500)(A/F,8\%,4)$$
$$= [10{,}000 + (2300)(.9259) + (2500)(.8573) + (3300)(.7938)](.3019) + (2300)(.2219)$$
$$= 5610.17$$

SO SELL AT END OF 3RD YEAR

29

$$EUAC(1) = 6000(A/P,10\%,1) + 400 = 6000(1.100) + 400 = 7000$$
$$EUAC(2) = 6000(A/P,10\%,2) + 400 + 100(A/G,10\%,2) = 6000(.5762) + 400 + 100(.4762) = 3904.82$$
$$EUAC(3) = 6000(.4021) + 400 + 100(.9366) = 2906.26$$

↓

$$EUAC(14) = 6000(.1357) + 400 + 100(4.9955) = 1713.75$$
$$EUAC(15) = 6000(.1315) + 400 + 100(5.2789) = 1716.89$$

END OF 14TH YEAR

30 SL $\frac{500{,}000 - 100{,}000}{25} = 16000$

SOYD $T = \frac{1}{2}(25)(26) = 325$

$$D_1 = \frac{25}{325}(500{,}000-100{,}000) = 30769.23$$
$$D_2 = \frac{24}{325}(400{,}000) = 29538.46$$
$$D_3 = \frac{23}{325}(400{,}000) = 28307.69$$

DDB

$$D_1 = \frac{2}{25}(500{,}000) = 40{,}000$$
$$D_2 = \frac{2}{25}(500{,}000 - 40{,}000) = 36{,}800$$
$$D_3 = \frac{2}{25}(500{,}000 - 40{,}000 - 36800) = 33{,}856$$

31 SL $\frac{12000 - 2000}{6} = 1666.67$

SOYD $T = \frac{1}{2}(6)(7) = 21$

$$D_5 = \frac{2}{21}(10{,}000) = 952.38$$

DDB TENTATIVELY calculate {EQN 2.21}

$$D_5 = \left(\frac{2}{6}\right)(12000)\left(1 - \frac{2}{6}\right)^4 = 790$$

CHECK THAT THE BOOK VALUE (EQN. 2.22) IS NOT LESS THAN \$2000

$$BV_5 = 12000\left(1 - \frac{2}{6}\right)^5 = 1580.24$$

SINCE $BV < 2000$, D_5 IS TOO MUCH. THE MAXIMUM DEPRECIATION ALLOWED IS $BV_4 - 2000$

$$BV_4 = 12000\left(1 - \frac{2}{6}\right)^4 = 2370.37$$
$$D_5 = 2370.37 - 2000 = 370.37$$

32 SL $D = \frac{2500 - 1100}{6} = 233.33/\text{YEAR}$

$$BV_1 = 2500 - 233.33 = 2266.67$$
$$BV_2 = 2266.67 - 233.33 = 2033.34$$
$$BV_3 = 2033.34 - 233.33 = 1800.01$$
$$BV_4 = 1800.01 - 233.33 = 1566.68$$
$$BV_5 = 1566.68 - 233.33 = 1333.35$$
$$BV_6 = 1333.35 - 233.33 = 1100.02$$

SOYD $T = \frac{1}{2}(6)(7) = 21$

$$D_1 = \frac{6}{21}(2500-1100) = 400$$
$$BV_1 = 2500 - 400 = 2100$$
$$D_2 = \frac{5}{21}(1400) = 333.33$$
$$BV_2 = 2100 - 333.33 = 1766.67$$
$$D_3 = \frac{4}{21}(1400) = 266.67$$
$$BV_3 = 1766.67 - 266.67 = 1500$$
$$D_4 = \frac{3}{21}(1400) = 200$$
$$BV_4 = 1500 - 200 = 1300$$

PROBLEM 32 CONTINUED

$D_5 = \frac{2}{21}(1400) = 133.33$

$BV_5 = 1300 - 133.33 = 1166.67$

$D_6 = \frac{1}{21}(1400) = 66.67$

$BV_6 = 1166.67 = 1100$

DDB

USING EQN 2.22, $d = \frac{2}{6} = \frac{1}{3}$

$BV_1 = 2500(1-\frac{1}{3})^1 = 1666.67$

$D_1 = 2500 - 1666.67 = 833.33$

$BV_2 = 2500(1-1/3)^2 = 1111.11$

$D_2 = 1666.67 - 1111.11 = 555.56$

SINCE BOOK VALUE CANNOT DROP BELOW 1100,

$BV_3 = BV_4 = BV_5 = BV_6 = 1100$

$D_3 = 1111.11 - 1100 = 11.11$

SINKING FUND

"BASIS" $= (2500-1100)(A/F, 6\%, 6)$

$= (1400)(0.1434) = 200.76$

FROM EQN 2.13,

$D_1 = (200.76)(1) = 200.76$

$BV_1 = 2500 - 200.76 = 2299.24$

$D_2 = (200.76)(F/P, 6\%, 1) = (200.76)(1.06) = 212.81$

$BV_2 = 2299.24 - 212.81 = 2086.43$

$D_3 = (200.76)(1.1236) = 225.57$

$BV_3 = 2086.43 - 225.57 = 1860.86$

$D_4 = (200.76)(1.1910) = 239.11$

$BV_4 = 1860.86 - 239.11 = 1621.75$

$D_5 = (200.76)(1.2625) = 253.46$

$BV_5 = 1621.75 - 253.46 = 1368.29$

$D_6 = (200.76)(1.3382) = 268.66$

$BV_6 = 1368.29 - 268.66 = 1099.63$

33 SL $DR = (.53)(233.33)(P/A, 6\%, 6)$

$= (.53)(233.33)(4.9173) = 608.10$

SOYD $D_1 = 400$

$G = \frac{1}{21}(2500 - 1100) = 66.67$

$DR = .53[(400)(P/A, 6\%, 6) - 66.67(P/G, 6\%, 6)]$

$= .53[(400)(4.9173) - (66.67)(11.4594)]$

$= 637.55$

DDB

EQUATION 2.20 CANNOT BE USED BECAUSE THE BOOK VALUE DOES NOT DECREASE BELOW 1100.

$DR = .53[(833.33)(P/F, 6\%, 1) + (555.56)(P/F, 6\%, 2) + (11.11)(P/F, 6\%, 3)]$

$= .53[(833.33)(.9434) + (555.56)(.8900) + (11.11)(.8396)]$

$= 683.67$

SINKING FUND

$DR = (.53)(200.76)(P/A, 6\%, 6)$

$= (.53)(200.76)(4.9173) = 523.21$

34 $D = \frac{80{,}000}{25} = 3200$

$P = -80{,}000 + (22{,}500 - 12{,}000)(1-.53)(P/A, 10\%, 25) + 3200(.53)(P/A, 10\%, 25)$

$= -80{,}000 + (10{,}500)(.47)(9.0770) + 3200(.53)(9.0770) = -19{,}810.41$

35 $T = \frac{1}{2}(25)(26) = 325$

$D_1 = \frac{25}{325}(80{,}000) = 6153.85$

$G = \frac{1}{325}(80{,}000) = 246.15$

$P = -80{,}000 + (22{,}500 - 12{,}000)(.47)(P/A, 10\%, 25) + 6153.85(.53)(P/A, 10\%, 25) - 246.15(.53)(P/G, 10\%, 25)$

$= -80{,}000 + (10{,}500)(.47)(9.0770) + 6153.85(.53)(9.0770) - 246.15(.53)(67.6964)$

$= -14{,}431.64$

36 SL $D = \frac{2000}{4} = 500$

$P = -2000 + 1200(1-.30)(P/A, 8\%, 4) + 500(.30)(P/A, 8\%, 4)$

$= -2000 + 1200(.70)(3.3121) + 500(.30)(3.3121) = 1278.98$

SOYD $T = \frac{1}{2}(4)(5) = 10$

$D_1 = \frac{4}{10}(2000) = 800$

$G = \frac{1}{10}(2000) = 200$

$P = -2000 + 1200(1-.30)(P/A, 8\%, 4) + 800(.30)(P/A, 8\%, 4) - 200(.30)(P/G, 8\%, 4)$

$= -2000 + 1200(.70)(3.3121) + 800(.30)(3.3121) - 200(.30)(4.6501)$

$= 1298.06$

37 $(1+.015)^{12} = 1.1956$

19.56%

38 SET THIS UP AS A RATE OF RETURN PROBLEM.

$100 = 9.46(P/A,?,12)$

$(P/A,?,12) = \frac{100}{9.46} = 10.57$

SEARCHING THE TABLES, ? = 2.00% PER MONTH

$i = (1+.02)^{12} - 1 = .2682$

OR 26.82%

39 $2000 = 89.30(P/A,?,30)$

$(P/A,?,30) = 22.40$

? = 2.00 PER MONTH

$i = (1.02)^{12} - 1 = .2682$

OR 26.82%

40 VARIABLE COST = $\frac{4000-3400}{360-240} = 5$

FIXED COST = $4000 - 5(360) = 2200$

AVERAGE COST OVER 240 UNITS

$= \frac{3400}{240} = 14.17$

NOT COUNTING THE FIXED COST,

PROFIT = $(249-240+1)(10.47-5) = 54.70$

COUNTING THE FIXED COST

PROFIT = $54.70 - \frac{2200}{249}(10) = -33.65$

41 FIND THE ANNUAL COST AND DIVIDE BY THE ANNUAL PRODUCTION

ANNUAL COST OF MACHINE =

$40{,}000(A/P,10\%,8) - 5000(A/F,10\%,8)$

$= 40{,}000(.1874) - 5000(.0874) = 7059$

HOURS USED PER YEAR

$= \left(\frac{48}{60}\right)2000 = 1600$

COST OF OPERATION

$= 800 + 1600(4.90+1.15) = 10480$

UNIT COST = $\frac{7059+10480}{2000} = 8.77$

42 QTY = $.62(700{,}000) = 434{,}000$

TOTAL COST = $190{,}000 + .348(434{,}000)$

$= 341{,}032$

PROFIT = $430{,}000 - 341{,}032 = 88968$

INCREMENTAL INCOME,

$= \frac{430{,}000}{434{,}000} = .991$

SO $190{,}000 + .348X = .991X$

$X = 295{,}490$ UNITS

43 SAVINGS PER YEAR

$= (3500)(.06) = 210$

$700 = (210-40)(P/A,10\%,?)$

$(P/A,10\%,?) = \frac{700}{170} = 4.118$

SEARCHING THE 10% TABLE, THE PAYBACK IS BETWEEN 5 AND 6 YEARS. N CAN BE FOUND EXACTLY (N = 5.58) BY EXTRACTING IT FROM THE (P/A) FORMULA, BUT THIS TAKES A LOT OF TIME.

44 $200(A/P,5\%,?) + 1.21(4000)$

$= 3600(A/P,5\%,?) + .75(4000)$

$(3600-200)(A/P,5\%,?) = (1.21-.75)(4000)$

$(A/P,5\%,?) = .5412$

? IS CLOSE TO 2 YEARS

E-I-T HOMEWORK SOLUTIONS: FLUID STATICS

1 PSIA = $P_{ATMOS} - P_{VACUUM}$

$= (29)\ \text{IN}\ (.491)\frac{\text{LBM}}{\text{IN}^3} - 9.5 = 4.739$ PSIA

2 $C = 3.4\ \text{EE-6}\ \ 1/\text{PSI}$

$\rho = 62.43\ \text{LBM/FT}^3$ {PAGE 4-32}

$E = 1/C = 2.94\ \text{EE5 PSI}$

FROM EQN 4.20 ON PAGE 4-5,

$c = \sqrt{Eg/\rho}$

$= \sqrt{\dfrac{(2.94\ \text{EE5})\frac{\text{LBF}}{\text{IN}^2}(144)\frac{\text{IN}^2}{\text{FT}^2}(32.2)\ \text{FT/SEC}^2}{62.43\ \text{LBF/FT}^3}}$

$= 4672.9$ FT/SEC

3 ASSUME $D_i = 4.00''$

$A = \frac{\pi}{4}(4.0)^2 = 12.566\ \text{IN}^2$

$F_p = PA = (900)\frac{\text{LBF}}{\text{IN}^2}(12.566)\ \text{IN}^2$

$= 11309.4$ LBF

$F_b = \frac{F_p}{3} = 3769.8$ LBF

4 FROM PAGE 4-32,

$E_{60^\circ} = 311\ \text{EE3 LBF/IN}^2$

FROM EQN 4.18 ON PAGE 4-5

$C = 1/E = 1/(311\ \text{EE3}) = 3.215\ \text{EE-6 IN}^2/\text{LBF}$

FROM EQN 4.17,

$\Delta V = C\Delta P V = (3.215\ \text{EE-6})\frac{\text{IN}^2}{\text{LBF}}(5000)\frac{\text{LBF}}{\text{IN}^2}(4.9)\ \text{FT}^3$

$= 7.877\ \text{EE-2 FT}^3$

5 FROM EQN 4.15 ON PAGE 4-5

$h = \frac{4(\cos\beta)\tau}{\rho d}$

$\beta = 0^\circ$ {TABLE 4.5}

$\tau = \frac{5.00\ \text{EE-3}}{12} = 4.16\ \text{EE-4 LBF/IN}$ (TABLE 4.5)

$\rho = 62.3\ \text{LBM/FT}^3$ {PAGE 4-32}

$= .0361\ \text{LBM/IN}^3$

$d = 0.1''$

$h = \frac{(4)(\cos 0^\circ)(4.16\ \text{EE-4})}{(.0361)(.1)} = .46''$

6
a) .00362 PSF {TABLE 4.3}
b) 122.4 PSF {TABLE 4.3}
c) 48.9 PSF {TABLE 4.3}
d) .9492 PSI {PAGE 6-30}
e) 14.696 PSI {PAGE 6-30}

7 a) $\rho_{Hg} = .491\ \text{LBM/IN}^3$

$h = \frac{p}{\rho} = \frac{(13.9)\ \text{LBF/IN}^2}{(.491)\ \text{LBM/IN}^3} = 28.31$ IN

THIS ASSUMES NEGLIGIBLE VAPOR PRESSURE

b) $\rho_w \approx .0361\ \text{LBM/IN}^3$

$P_{v,70^\circ} = .3631$ PSI {PAGE 6-30}

$h = \frac{P_a - P_v}{\rho} = \frac{13.9 - .3631}{.0361} = 375''$

c) $(SG)_{ALC} = .789$ {PAGE 4-33}

$\rho = (SG)(\rho_{WATER}) = (.789)(.0361) = .0285$

$P_v = \frac{(122.4)}{(144)} = .85\ \text{LBF/IN}^2$

$h = \frac{13.9 - .85}{.0285} = 457.9''$

8 THE DENSITY OF SEA WATER IS ABOUT $64.0\ \text{LBM/FT}^3$

$p = \rho h = (64.0)(8000) = 512{,}000\ \text{LBF/FT}^2$

9 IF THE 2 CHAMBERS DID NOT COMMUNICATE THE PRESSURE UNDER THE LEFT CHAMBER WOULD BE

$p = \sum \rho h = (3)(.7)(62.4) + (6)(1.3)(62.4) = 617.76$ PSF

SIMILARLY THE PRESSURE UNDER THE RIGHT CHAMBER WOULD BE

$(3)(1)(62.4) + (4)(1.3)(62.4) = 511.68$

SINCE THE 2 CHAMBERS DO COMMUNICATE, THE PRESSURE MUST BE 617.76 PSF EVERYWHERE ON THE BOTTOM. WORKING UPWARDS ON THE RIGHT HAND SIDE,

$P_A = 617.76 - (4)(1.3)(62.4) - (3)(1)(62.4)$

$= 617.76 - 511.68 = 106.08$ PSF

$= .737$ PSIG

10 THE PRESSURES AT VARIOUS DEPTHS ARE

DEPTH	PRESSURE
0	0
2	(2)(55) = 110 PSF
4	110 + 2(62.4) = 234.8
6	234.8 + 2(62.4) = 359.6

THE AVERAGE PRESSURE ON THE INCLINED SURFACE IS {FROM EQN 4-30}

$\frac{1}{2}(234.8 + 359.6) = 297.2$ PSF

$F_x = 5[(2)(297.2)] = 2972$ LBF

$F_y = 5[(2)(2)(55) + (2)(2)(62.4) + \frac{1}{2}(2)(2)(62.4)]$

$= 2972$ LBF

11 $\rho_{AIR} = P/RT = \dfrac{(26+14.7)(144)}{(53.3)(460+70)} = .2075\ \text{LBM/FT}^3$

$P_{BOTTOM} = P_a + P_g + \sum \rho h$

$= 14.7 + 26.0 + \dfrac{(62.4)(9.3)+(62.4)(.71)(6.5)+(.2075)(5)}{144}$

$= 46.737$ PSIA

12 $\bar{p} = \frac{1}{2}(5)(62.4) = 156$ PSF

$A = (5)(10) = 50$

$F = \bar{p}A = (156)(50) = 7800$ LBF

$d = 1 + (\frac{2}{3})5 = 4.33$ FT

13 THE PRESSURE AT THE BOTTOM OF THE LEFT CHAMBER IS

$p = \dfrac{F}{A} + \rho h = \dfrac{200\ \text{LBF}}{\frac{\pi}{4}(1)^2} + (.0361)(.7)(8)$

$= 254.85$ psig

WORKING UPWARDS TOWARD B

$P_B = 254.85 - (10)\,\text{FT}\,(12)\,\frac{\text{IN}}{\text{FT}}(.7)(.0361)\,\frac{\text{LBM}}{\text{IN}^3}$

$= 251.82$ psi

THE FORCE IS

$F_B = P_B A = (251.82)(\frac{\pi}{4})(12)^2 = 28480$ LBF

14 THE PRESSURE IN THE JAR IS THE SAME AS THE PRESSURE AT THE LOWER WATER SURFACE.

$p = \rho h = (.0361)(16) = .5776$ psig

15 $h_2 = 15/\sin 20° = 43.86$ (INCLINED)

$h_1 = 43.86 - 8 = 35.86$

$\bar{p} = (62.4)(\frac{1}{2})(35.86 + 43.86)(\sin 20°) = 850.7$ PSF

$F = \bar{p}A = (850.7)[4 \times 8] = 27{,}222.4$

FROM EQN 4.32,

$h_R = \frac{2}{3}\left[35.86 + 43.86 - \dfrac{(35.86)(43.86)}{35.86+43.86}\right]$

$= 40.0'$ (INCLINED)

16 a) $P_A = 0$

$P_C = (5)(62.4) = 312$ PSF

$\bar{p} = \frac{1}{2}(0+312) = 156$ PSF

$F_x = \bar{p}A = (156)(5)(1) = 780$ LBF

b) F_y IS THE WEIGHT OF THE WATER A-B-C

$WT = \rho V$

$= (62.4)\,\frac{\text{LBM}}{\text{FT}^3}\left[(2)(3) + \frac{\pi}{4}(3)^2\,\text{FT}^2\right](1)\ \text{FT}$

$= 815.5$ LBF

c) $R = \sqrt{(780)^2 + (815.5)^2} = 1128.5$ LBF

$\theta = \arctan\left(\dfrac{815.5}{780}\right) = 46.27°$ FROM HORIZ.

17 $P_B = (6)\,\text{FT}\,(62.4)\,\frac{\text{LBM}}{\text{FT}^3} = 374.4$ PSF

$P_C = (11)(62.4) = 686.4$ PSF

$\bar{p} = \frac{1}{2}(374.4 + 686.4) = 530.4$ PSF

$F_{LEFT\,SIDE} = pA = 530.4\,\frac{\text{LBF}}{\text{FT}^2}\left[(5)\,\text{FT}\,(1)\,\text{FT}\right] = 2652$ LBF

FROM EQN 4.32 THE RESULTANT IS LOCATED AT

$h_R = \frac{2}{3}\left[6 + 11 - \dfrac{(6)(11)}{6+11}\right]$

$= 8.75$ FROM TOP

OR 2.25 FROM BOTTOM

SIMILARLY FOR THE DOWNSTREAM WATER,

$P_C = (2)(62.4) = 124.8$

$\bar{p} = \frac{1}{2}(0 + 124.8) = 62.4$ PSF

$F_{RIGHT\,SIDE} = pA = (62.4)[(2)(1)] = 124.8$ LBF

THIS IS LOCATED $\frac{1}{3}(2) = .67$ FROM BOTTOM

NOW, DRAW A FREEBODY OF THE GATE

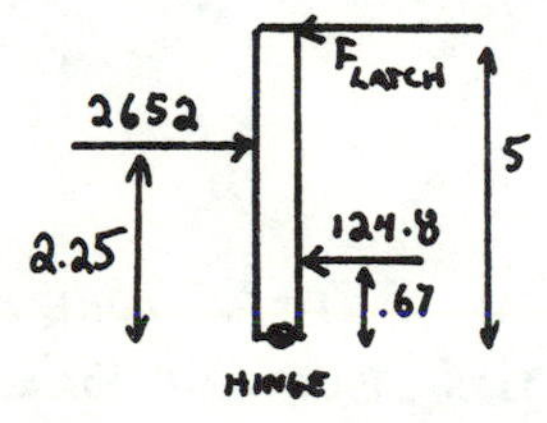

SUMMING MOMENTS ABOUT THE HINGE,

$\sum M \circlearrowleft = 0: (124.8)(.67) + (F_{LATCH})(5) - (2652)(2.25) = 0$

$F_{LATCH} = 1176.7$ LBF

18

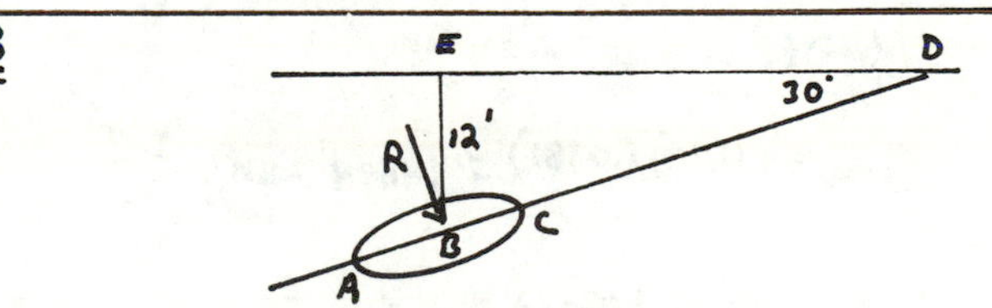

$h_A = 12 + 3.5(\sin 30°) = 13.75$ FT

$h_C = 12 - 3.5(\sin 30°) = 10.25$ FT

$A = \frac{\pi}{4}(7)^2 = 38.48$ FT^2

$P_A = (13.75)(62.4) = 858$ PSF

$P_C = (10.25)(62.4) = 639.6$ PSF

$\bar{p} = \frac{1}{2}(858 + 639.6) = 748.8$ PSF

$R = \bar{p}A = (748.8)(38.48) = 28813.8$ LBF, 60° FROM HORIZ

SINCE THIS IS NOT A RECTANGLE, EQUATION 4.35 MUST BE USED TO FIND h_R

$I_C = \frac{1}{4}\pi r^4$ {PAGE 9-19}

$= (\frac{1}{4})\pi(3.5)^4 = 117.86$ FT^4

h_C IS MEASURED PARALLEL TO DISK

$h_C = 12/\sin 30° = 24$ FT

{MORE}

PROBLEM 18 CONTINUED

THEN, h_R MEASURED PARALLEL TO DISK IS

$$h_R = 24 + \frac{117.86}{(3848)(24)} = 24.13$$

THIS IS A VERTICAL DEPTH OF

$$h = (24.13)\sin 30° = 12.06 \text{ FT}$$

19

$$\Delta P = (6)\text{ IN}(.491)\frac{\text{LBM}}{\text{IN}^3} - (8)\text{ IN}(.0361)\frac{\text{LBM}}{\text{IN}^3}$$
$$= 2.657 \text{ psig}$$

20

ASSUME THAT THERE IS WATER IN THE MANOMETER TUBES

$$\Delta p = 18(.491) - 18(.0361) = 8.188 \text{ psig}$$

21

$$p = 15(.491) + 30(.82)(.0361) = 8.253 \text{ psig}$$

22

THE BOX WEIGHT IS

$$[(12)^3 - (11.5)^3](7.7)(.0361)$$
$$= 57.6 \text{ LB}$$

THE BOX VOLUME IS 1 FT^3. WHEN SUBMERGED, THE BOX BUOYANT FORCE WILL BE 62.4 LBF. SINCE 62.4 > 57.6, THE BOX WILL FLOAT

23

$$F_{\text{BUOYANT}} = \text{WEIGHT}$$

$$\Sigma(\text{VOL})(\rho_{\text{WATER}}) = \text{WEIGHT}$$

$$\rho_{\text{LEAD}} = (11.34)(.0361) = .4094 \text{ LBM/IN}^3$$

$$[(8)\text{ IN}^3 + V_{\text{LEAD}}](.0361)\frac{\text{LBM}}{\text{IN}^3} =$$

$$= (8)\text{ IN}^3(.25)(.0361)\frac{\text{LBM}}{\text{IN}^3} + (V_{\text{LEAD}})(.4094)$$

$$.2888 + .0361 V_{\text{LEAD}} = .0722 + .4094 V_{\text{LEAD}}$$

$$V_{\text{LEAD}} = .58 \text{ IN}^3$$

$$\text{WT} = \rho V = (.4094)(.58) = .2375 \text{ LBM}$$

24

WT OF STONE DRY − BUOYANT FORCE = WT OF STONE SUBMERGED

$$19.9 - (V_{\text{STONE}})(62.4) = 12.4$$

$$V_{\text{STONE}} = .12 \text{ FT}^3$$

$$\rho = \frac{19.9}{.12} = 165.8 \text{ LBM/FT}^3$$

25

THE ADDITIONAL DISPLACED VOLUME IS

$$(4700)\text{ FT}^2\left(\frac{3}{12}\right)\text{FT} = 1175 \text{ FT}^3$$

$$F_{\text{BUOY}} = (1175)\text{ FT}^3(64.0)\frac{\text{LBM}}{\text{FT}^3} = 75200 \text{ LBM}$$

26

$$\rho_{\text{AIR}} = P/RT = \frac{(14.7)(144)}{(53.3)(460+55)} = .0771 \text{ LBM/FT}^3$$

$$F_{\text{BUOY}} = \text{WT OF AIR} = (.0771)\frac{\text{LBM}}{\text{FT}^3}(16{,}000)\text{ FT}^3$$
$$= 1233.6 \text{ LBF}$$

R FOR HYDROGEN = 766.8 (PAGE 6-12)

$$\rho_H = \frac{(14.7)(144)}{(766.8)(515)} = .00536 \text{ LBM/FT}^3$$

$$\text{WT OF HYDROGEN} = (.00536)(16000) = 85.76 \text{ LBM}$$

FOR IMPENDING LIFT-OFF,

$$\text{WT} = F_B$$

$$650 + \text{WT}_{\text{SAND}} + 85.76 = 1233.6$$

$$\text{WT}_{\text{SAND}} = 497.84 \text{ LBM}$$

27

LET X BE THE % OF SUBMERGED ICEBERG AND LET V BE THE VOLUME, THEN

$$\text{WEIGHT} = F_B$$

$$(V)(57.1) = (V)(X)(62.4)$$

$$X = .9151$$

$$\% \text{ VISIBLE} = 1 - .9151 = .0849$$
OR 8.49%

28

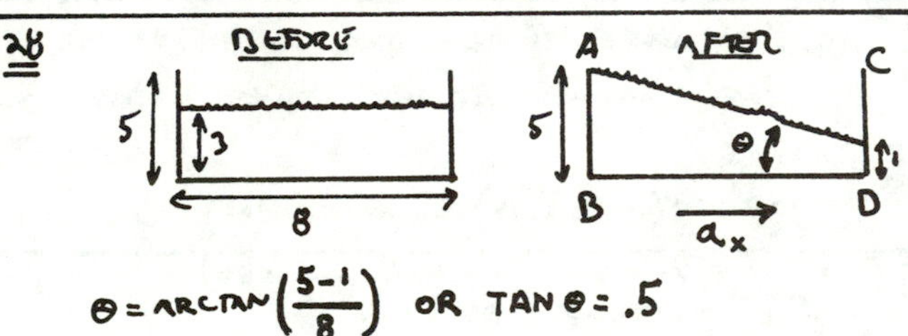

$$\theta = \arctan\left(\frac{5-1}{8}\right) \text{ OR } \tan\theta = .5$$

a) WORK WITH 1 FOOT WIDTH. FROM EQN 4.44,

$$\tan\theta = \frac{a_x}{g + a_y}$$

$$.5 = \frac{a_x}{32.2 + 0} \quad \text{OR} \quad a_x = 16.1 \text{ FT/SEC}^2$$

b)

$$P_A = 0$$
$$P_B = (5)(62.4) = 312 \text{ PSF}$$
$$\bar{p} = \tfrac{1}{2}(0 + 312) = 156 \text{ PSF}$$
$$F_{\text{REAR}} = PA = (156)(5)(8) = 6240 \text{ LBF}$$

SIMILARLY

$$P_C = 0$$
$$P_D = (1)(62.4) = 62.4$$
$$\bar{p} = \tfrac{1}{2}(0 + 62.4) = 31.2$$
$$F_{\text{FRONT}} = (31.2)(1)(8) = 249.6 \text{ LBF}$$

29

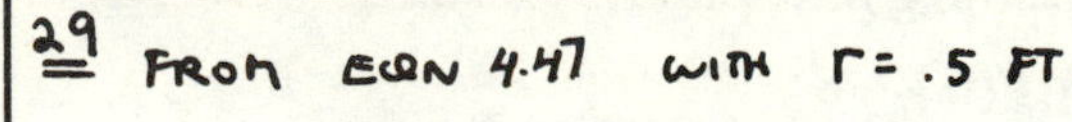

FROM EQN 4.47 WITH $r = .5$ FT

$$\omega = \frac{\sqrt{2gh}}{r} = \frac{\sqrt{(2)(32.2)h}}{.5} = 16.05\sqrt{h}$$

BUT h IS UNKNOWN. h IS NOT EQUAL TO 1 FOOT, SINCE AS THE WATER CLIMBS THE WALL, THE VERTEX WILL DROP. HOWEVER, THE VOLUME OF THE WATER IS CONSTANT.

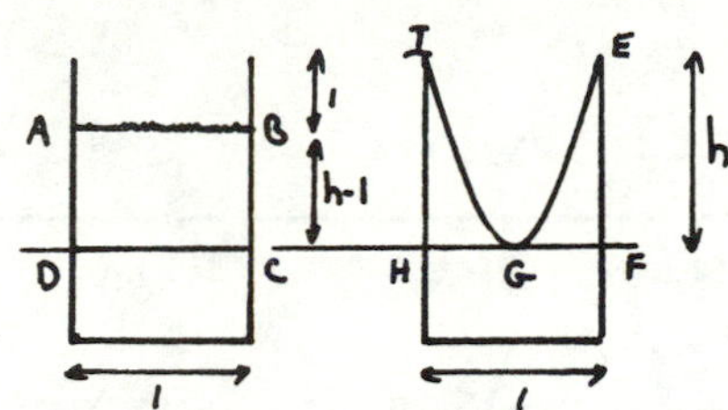

THE VOL (ABCD) = VOL (EFHI) − VOL (EGI)

USING PAGES 1-1 AND 1-2,

$$\frac{\pi}{4}(1)^2(h-1) = \frac{\pi}{4}(1)^2(h) - \frac{\pi}{8}(1)^2 h$$

$$h = 2$$

THEN $\omega = 16.05\sqrt{2} = 22.7$ RAD/SEC

$$\frac{22.7}{2\pi} = 3.61 \text{ REV/SEC}$$

30

FROM EQN 4.36 WITH $h_1 = 0$

$$\int \frac{dp}{\rho} = -h_2$$

$$T = 59°F = 519°R$$

$$\rho = p/RT = \frac{p}{(53.3)(519 - .0035h_2)}$$

$$\int \frac{(53.3)(519 - .0035h_2)}{p}\, dp = -h_2$$

$$\ln(p_2) - \ln(p_1) = \frac{-h_2}{(53.3)(519 - .0035h_2)}$$

$$\ln p_2 = \frac{-35{,}000}{(53.3)(519 - .0035(35000))} + \ln\left[(14.7)(144)\right]$$

$$= 6.002$$

$$p_2 = 404.04 \text{ PSF}$$

$$= 2.806 \text{ PSIA}$$

31

a) $\rho = p/RT = \dfrac{(14.7)(144)}{(53.3)(460+60)} = .0764 \text{ LBM/FT}^3$

$$p = \frac{(14.7)\frac{\text{LBF}}{\text{IN}^2}(144)\frac{\text{IN}^2}{\text{FT}^2} - (12000)\text{FT}\,(.0764)\frac{\text{LBM}}{\text{FT}^3}}{144 \text{ IN}^2/\text{FT}^2}$$

$$= 8.33 \text{ PSIA}$$

b) FROM EQN 4.38

$$p_2 = 14.7 \exp\left[\frac{0 - 12000}{53.3(520)}\right] = 9.53 \text{ PSIA}$$

c) AN ADIABATIC PROCESS IS A POLYTROPIC PROCESS WITH $N = K$. FROM EQN 4.40

$$p_2 = 14.7\left[1 - \left(\frac{1.4-1}{1.4}\right)\left(\frac{12000 - 0}{(53.3)(520)}\right)\right]^{\frac{1.4}{1.4-1}} = 14.7(.8763)^{3.5}$$

$$= 9.26 \text{ PSIA}$$

E-I-T HOMEWORK SOLUTIONS: FLUID DYNAMICS

1 FROM EQN 4.53

a) $\frac{v^2}{2g} = \frac{(15)^2}{(2)(32.2)} = 3.49$ FT

b) $\frac{(15)^2}{(2)(32.2)} = 3.49$ FT

2 FROM PAGE 4-32

$\nu_{AIR, 250°} = 27.3$ EE-5 FT²/SEC

$N_{Re} = \frac{Dv}{\nu} = \frac{(2/12) \text{ FT } (120) \text{ FT/SEC}}{(27.3 \text{ EE-5}) \text{ FT}^2/\text{SEC}} = 7.326$ EE4

3 FROM PAGE 4-32

$\nu = .739$ EE-5 FT²/SEC

$N_{Re} = \frac{Dv}{\nu} = \frac{(4/12) \text{ FT } (20) \text{ FT/SEC}}{.739 \text{ EE-5 FT}^2/\text{SEC}} = 9.02$ EE5

4 $N_{Re} = \frac{Dv}{\nu} = \frac{(3/12)(10)}{.005} = 500$

SINCE $N_{Re} < 2000$, THE FLOW IS LAMINAR

5 a) FROM THIS DIAGRAM, IT IS APPARENT THAT $\phi = 60°$

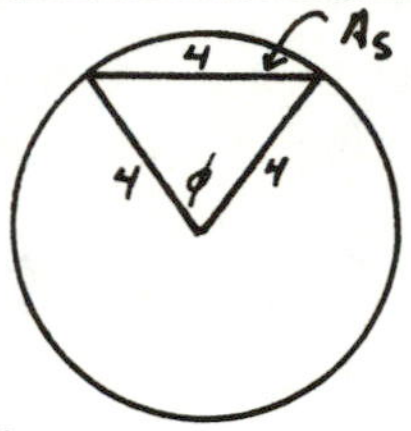

SO $\phi = \frac{60}{360}(2\pi) = 1.047$ RADIANS

FROM PAGE 1-2

$A_s = \frac{1}{2}(r^2)(\phi - \sin\phi)$

$= (\frac{1}{2})(4)^2(1.047 - \sin(1.047)) = 1.449$

$s = r\phi = (4)(1.047) = 4.188$

THE WETTED PERIMETER IS

$2\pi r - s = 2\pi(4) - 4.188 = 20.945$

THE AREA IN FLOW IS

$\pi r^2 - A_s = \pi(4)^2 - 1.449 = 48.814$

FROM EQN 4.65,

$r_H = \frac{48.816}{20.945} = 2.33$

b) $s = \frac{7}{\sin 20°} = 20.47$

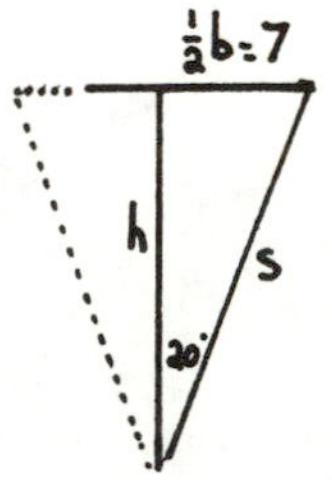

$h = \frac{7}{\tan 20°} = 19.23$

THE WETTED PERIMETER IS

$2s = 2(20.47) = 40.94$

THE AREA IN FLOW IS

$\frac{1}{2}bh = (\frac{1}{2})(14)(19.23) = 134.61$

THEN, FROM EQN 4.65

$r_H = \frac{134.61}{40.94} = 3.29$

c) FROM PAGE 1-2

$a = \frac{1}{2}(16) = 8$

$b = \frac{1}{2}(12) = 6$

AREA IN FLOW IS

$\frac{1}{2}(\pi ab) = (\frac{1}{2})(\pi)(8)(6) = 75.4$

THE WETTED PERIMETER IS

$\frac{1}{2}(2\pi)\sqrt{\frac{1}{2}(a^2+b^2)} = \frac{1}{2}(2\pi)\sqrt{\frac{1}{2}(8^2+6^2)}$

$= 22.21$

$r_H = \frac{75.4}{22.21} = 3.39$

6 FROM EQN 4.68

$A_1 v_1 = A_2 v_2$

$(\frac{\pi}{4})(2)^2(12) = (\frac{\pi}{4})(5)^2(v_B)$

$v_B = 1.92$ FPS

7 FROM EQN 4.66

$\rho_1 A_1 v_1 = \rho_2 A_2 v_2$

$(.065) \frac{\text{LBM}}{\text{FT}^3} (\frac{\pi}{4})(\frac{4}{12})^2 \text{FT}^2 (50) \frac{\text{FT}}{\text{SEC}} =$

$= (\rho_B)(\frac{\pi}{4})(\frac{7}{12})^2(20)$

$\rho_B = .05306$ LBM/FT³

8 $\dot{m} = \rho Q = (62.4) \frac{\text{LBM}}{\text{FT}^3} (1000) \frac{\text{FT}^3}{\text{SEC}}$

$= 6.24$ EE4 $\frac{\text{LBM}}{\text{SEC}}$

THE HEAD AVAILABLE FOR WORK IS

$\Delta h = 625 - 58 = 567$ FT

THE WORK DONE IS

$W = \eta \dot{m} \Delta h = (.89)(6.24 \text{ EE4}) \frac{\text{LBM}}{\text{SEC}} (567)$ FT

$= 3.149$ EE7 $\frac{\text{FT-LBF}}{\text{SEC}}$

FROM PAGE 7-1,

$(3.149 \text{ EE7}) \frac{\text{FT-LBF}}{\text{SEC}} (1.356 \text{ EE-3}) \frac{\text{KW-SEC}}{\text{FT-LBF}} =$

$= 4.27$ EE4 KW

9 AT A

$v_A = \frac{Q}{A} = \frac{2.5 \text{ FT}^3/\text{SEC}}{\frac{(12.5) \text{ IN}^2}{144 \text{ IN}^2/\text{FT}^2}} = 28.8$ FT/SEC

THE VELOCITY HEAD AT A IS

$h_{v,A} = \frac{v^2}{2g} = \frac{(28.8)^2}{(2)(32.2)} = 12.88$ FT

THE STATIC HEAD AT A IS

$h_{s,A} = \frac{(10) \text{PSIG} (144)}{62.4} + 1.5 = 24.58$ FT

THE POTENTIAL HEAD IS ZERO IF A IS CHOSEN AS THE DATUM

(MORE)

PROBLEM 9 CONTINUED

AT B

THE STATIC HEAD AT B IS

$$\frac{\left((2.5)\text{ IN}(.491)\frac{\text{LBM}}{\text{IN}^3} - (12)(.0361)\frac{\text{LBM}}{\text{IN}^3}\right)144}{62.4}$$

$= 1.83$ FT

THE POTENTIAL HEAD IS 10 FT ABOVE A

ASSUMING FRICTIONLESS FLOW, THE TOTAL HEAD IS CONSTANT

$$12.88 + 24.58 + 0 = \frac{V_B^2}{(2)(32.2)} + 1.83 + 10$$

$V_B = 40.63$ FT/SEC

10 FROM PAGE 4-32 FOR 100°F AIR

$\nu = 18 \cdot EE\text{-}5$

SO $N_{Re} = \frac{DV}{\nu} = \frac{(12/12)(14)}{18\ EE\text{-}5} = 7.78\ EE4$

FROM P. 4-17 $\epsilon = .004$ FT

SO $\frac{\epsilon}{D} = \frac{.004'}{\left(\frac{12}{12}\right)} = .004$

FROM PAGE 4-17, $f = .0295$

FROM EQN 4.71,

$$h_f = \frac{(.0295)(100)\text{ FT }(14)^2\text{ FT}^2/\text{SEC}^2}{(2)(12/12)\text{ FT }(32.2)\text{ FT/SEC}^2} = 8.98\text{ FT}$$

11 FROM PAGE 4-32

$\nu = 1.217\ EE\text{-}5$ FT²/SEC

$N_{Re} = \frac{DV}{\nu} = \frac{(2/12)(4)}{1.217\ EE\text{-}5} = 5.48\ EE4$

FROM PAGE 4-17, $\epsilon = .0005$ FT

$\frac{\epsilon}{D} = \frac{.0005}{(2/12)} = .003$

FROM PAGE 4-17, $f = .0282$

$$h_f = \frac{(.0282)(1000)(4)^2}{(2)(2/12)(32.2)} = 42.04$$

12 750 gpm $= (750)(.002228) = 1.671$ FT³/SEC

THE VELOCITY IS

$$v = \frac{Q}{A} = \frac{1.671}{\left(\frac{\pi}{4}\right)\left(\frac{6}{12}\right)^2} = 8.51\text{ FT/SEC}$$

AT A

THE VELOCITY IS THE SAME AS AT B, SO MAY BE NEGLECTED.

THE GAGE PRESSURE HEAD IS h_A.

THE GRAVITATIONAL HEAD IS ZERO.

AT B

THE PRESSURE HEAD IS

$$\frac{(50)\frac{\text{LBM}}{\text{IN}^2}(144)\text{ IN}^2/\text{FT}^2}{(.9)(62.4)\frac{\text{LBM}}{\text{FT}^3}} = 128.21\text{ FT}$$

THE GRAVITATIONAL HEAD IS 60 FT

THE FRICTION BETWEEN POINTS A AND B IS

$$h_f = \frac{(.03)(3000)(8.51)^2}{(2)(6/12)(32.2)} = 202.42\text{ FT}$$

FROM EQN 4.70,

$$h_A + h_v + 0 - 202.42 = 128.21 + h_v + 60$$

SO $h_A = 390.63$ FT

$$p = \rho h = \frac{(.9)(62.4)(390.63)}{144} = 152.35\text{ psig}$$

13 $h_v = \frac{v^2}{2g} = \frac{(15)^2}{(2)(32.2)} = 3.494$

$h_f = (.8)(3.494) = 2.8$ FT

14

300 FT

FOR 70°F WATER, $\nu = 1.059\ EE\text{-}5$

$N_{Re} = \frac{DV}{\nu} = \frac{(12/12)(10)}{(1.059\ EE\text{-}5)} = 9.44\ EE5$

FROM PAGE 4-17,

$\epsilon = .0002$ FT

$\frac{\epsilon}{D} = \frac{.0002}{(12/12)} = .0002$

FROM PAGE 4-17, $f = .0147$

$$h_f = \frac{(.0147)(300)(10)^2}{(2)(12/12)(32.2)} = 6.85\text{ FT}$$

THE VELOCITY HEAD IS

$$\frac{v^2}{2g} = \frac{(10)^2}{(2)(32.2)} = 1.55\text{ FT}$$

$h_{MINOR} = .5(1.55) = .78$ FT

WHEN THE FLUID ENTERS THE SECOND TANK, ALL OF THE VELOCITY HEAD WILL BE LOST, SO THE TOTAL LOSS IS

$6.85 + .78 + 1.55 = 9.18$ FT

15 USE TABLE 4.9.

$L_e = 13.0 + 110 = 123$ FT

THE TOTAL LENGTH INCLUDING FITTINGS IS

$L = 150 + 123 = 273$

(MORE)

PROBLEM 15 CONTINUED

FOR WATER FROM PAGE 4-32

$\nu = 1.059$ EE-5

$N_{Re} = \frac{VD}{\nu} = \frac{(12)(4/12)}{1.059 \text{ EE-5}} = 3.78$ EE5

FROM PAGE 4-17 $\epsilon = .0002$

$\frac{\epsilon}{D} = \frac{.0002}{(4/12)} \approx .0006$

FROM PAGE 4-17 $f = .0185$

$h_f = \frac{(.0185)(273)(12)^2}{(2)(4/12)(32.2)} = 33.88$ ft

16

AT A $z = 100$, $v = 0$, $p = 0$

AT B $z = 0$

$h_v = \frac{v^2}{2g} = .0155\, v^2$

$p = 0$

$h_f = \frac{(.02)(2000)(v^2)}{(2)(3/12)(32.2)} = 2.48\, v^2$

FROM CONSERVATION OF ENERGY,

$100 = (.0155 + 2.48)\, v^2$

$v = 6.33$ FT/SEC FOR GRAVITY FLOW

NOW, IF THE FLOW IS DOUBLED BY A PUMP,

$V_2 = 2V_1 = (2)(6.33) = 12.66$ FT/SEC

$\dot{m} = Q\rho = \left(\frac{\pi}{4}\right)\left(\frac{3}{12}\right)^2(12.66)(62.4) = 38.78$ LBM/SEC

$h_v = \frac{(12.66)^2}{(2)(32.2)} = 2.49$ FT

$h_f = \frac{(.02)(2000)(12.66)^2}{(2)(3/12)(32.2)} = 398.2$ FT

THEN, FROM EQN 4.70

$0 + 0 + 100 + h_A - 398.2 = 0 + 2.49 + 0$

$h_A = 300.69$ FT

$W = \dot{m} h_A = (38.78)\frac{\text{LBM}}{\text{SEC}}(300.69)\text{ FT} = 11660.8 \frac{\text{FT-LBF}}{\text{SEC}}$

$HP = \frac{11660.8}{550} = 21.2$ HP

17

$Q = (2000)\text{ gpm}(.002228)\frac{\text{FT}^3}{\text{SEC-gpm}} = 4.456\ \text{FT}^3/\text{SEC}$

$\dot{m} = Q\rho = (4.456)(1.2)(62.4) = 333.67$ LBM/SEC

AT A

VELOCITY IS SAME AS AT B, SO IT IS NEGLECTED

$z = 0$

$h_p = \frac{[14.7 - (6)(.491)]\,144}{(62.4)(1.2)} = 22.60$

AT B

$z = 4$

$h_p = \frac{(20 + 14.7)\,144}{(62.4)(1.2)} = 66.73$ FT

THE HEAD ADDED BY THE PUMP IS

$h_A = 66.73 + 4 - 22.6 = 48.13$ FT

THE WORK DONE IS

$W = \dot{m} h_A = (333.67)\frac{\text{LBM}}{\text{SEC}}(48.13)\text{ FT}$

$= 16059.5 \frac{\text{FT-LBF}}{\text{SEC}}$

$HP = \frac{16059.5}{550} = 29.19$

18

$Q = (100)\text{ gpm}(.002228)\frac{\text{FT}^3}{\text{SEC-gpm}} = .2228\ \text{FT}^3/\text{SEC}$

$\dot{m} = Q\rho = (.2228)\ \text{FT}^3/\text{SEC}\,(.85)(62.4)\ \text{LBM}/\text{FT}^3$

$= 11.817$ LBM/SEC

AT A $h_{static} = \frac{(5 + 14.7)\,144}{(.85)(62.4)} = 53.48$ FT

$V_A = \frac{Q}{A} = \frac{.2228}{\left(\frac{\pi}{4}\right)\left(\frac{3}{12}\right)^2} = 4.539$ FT/SEC

$h_{velocity} = v^2/2g = \frac{(4.539)^2}{(2)(32.2)} = .32$ FT

AT B

$h_{static} = \frac{(100 + 14.7)\,144}{(.85)(62.4)} = 311.4$ FT

$V_B = \frac{Q}{A} = \frac{.2228}{\left(\frac{\pi}{4}\right)\left(\frac{2}{12}\right)^2} = 10.21$ FT/SEC

$h_{velocity} = \frac{(10.21)^2}{(2)(32.2)} = 1.62$ FT

FROM EQN 4.70

$h_{ADDED} = 311.4 + 1.62 - 53.48 - .32 = 259.22$ FT

$W = (h_{ADDED})\dot{m} = (259.22)\text{ FT}(11.817)\text{ LBM/SEC}$

$= 3063.2 \frac{\text{FT-LBF}}{\text{SEC}}$

$HP = W/550 = \frac{3063.2}{550} = 5.57$

19 FROM EQN 4.88 WITH $C_v = 1$, THE DISCHARGE VELOCITY IS

$V_0 = V_x = \sqrt{2g(20 - y)}$

FROM EQN 4.93

$t = \sqrt{2y/g}$

SO

$x = \sqrt{2g(20-y)(2y/g)} = \sqrt{4y(20-y)}$

(MORE)

PROBLEM 19 CONTINUED

MAXIMIZING x REQUIRES $\frac{dx}{dy} = 0$

$$\frac{dx}{dy} = \frac{1}{2}\left(4y(20-y)\right)^{-\frac{1}{2}}(4)\left[y(-1) + (1)(20-y)\right]$$

$$= \frac{20-2y}{\sqrt{(20-y)y}}$$

IF $\frac{dx}{dy} = 0$, THE NUMERATOR MUST BE ZERO, SO $y = 10$

20 $A_o = \frac{\pi}{4}\left(\frac{1}{12}\right)^2 = .005454 \text{ FT}^2$

FROM EQN 4.88

$$V_{IDEAL} = \sqrt{2gh} = \sqrt{2(32.2)(2.1)} = 11.63 \text{ FT/SEC}$$

THE IDEAL WT OF DISCHARGE OVER 90 SECONDS IS

$$(W)_{IDEAL} = (V_{IDEAL})A_o \rho t = (11.63)(.005454)(62.4)(90) = 356.2$$

$$C_D = \frac{W_{ACTUAL}}{W_{IDEAL}} = \frac{228}{356.2} = .64$$

FROM EQN 4.92

$$C_V = \frac{x}{2\sqrt{hy}} = \frac{4}{(2)\sqrt{(2.1)(2)}} = .976$$

THEN, SINCE $C_D = C_c C_V$

$$C_c = .64/.976 = .656$$

21 $A_o = \frac{\pi}{4}\left(\frac{4}{12}\right)^2 = .08727 \text{ FT}^2$

$A_t = \frac{\pi}{4}(20)^2 = 314.16 \text{ FT}^2$

FROM EQN 4.96 THE TIME TO DROP FROM 40 FT TO 20 FT IS

20

40

A_o

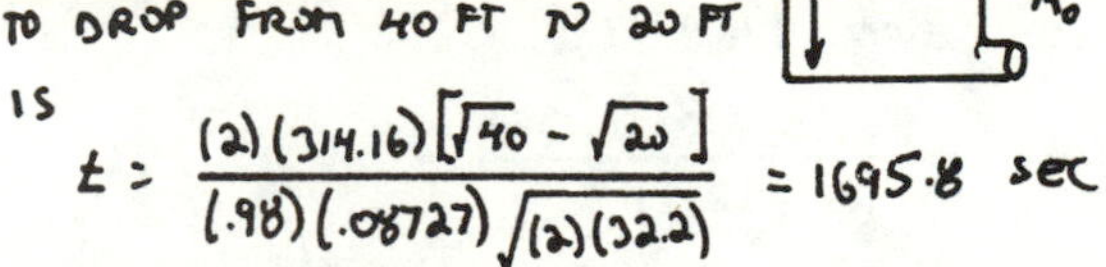

$$t = \frac{(2)(314.16)\left[\sqrt{40} - \sqrt{20}\right]}{(.98)(.08727)\sqrt{(2)(32.2)}} = 1695.8 \text{ SEC}$$

22 $C_d = 1.00$ {GIVEN}

$$F_{va} = \frac{1}{\sqrt{1 - (D_2/D_1)^4}} \quad \text{\{FROM EQN 4.129\}}$$

$$= \frac{1}{\sqrt{1 - (8/12)^4}} = 1.116$$

$$A_2 = \frac{\pi}{4}\left(\frac{8}{12}\right)^2 = .3491 \text{ FT}^2$$

$$P_1 - P_2 = \left[(.491)(4) - (.0361)(4)\right]144 = 262.0 \text{ LBF/FT}^2$$

THEN FROM EQN 4.128

$$Q = (1.116)(1)(.3491)\sqrt{\frac{(2)(32.2)(262.0)}{62.4}} = 6.406 \text{ FT}^3/\text{SEC}$$

23 DISREGARD THE COMPRESSIBILITY OF THE AIR.

$C_d = .98$

$$F_{VA} = \frac{1}{\sqrt{1 - (1.5/3)^4}} = 1.033$$

THEN $C_f = C_d F_{VA} = (.98)(1.033) = 1.012$

$$A_2 = \frac{\pi}{4}\left(\frac{1.5}{12}\right)^2 = .0123 \text{ FT}^2$$

$$P_1 = (30.05) \text{ IN Hg} + (.287) \text{ IN Hg} = 30.337 \text{ IN Hg}$$
$$= (30.337)(.491)(144) = 2144.95 \text{ LBF/FT}^2$$

$$\rho_1 = P_1/RT_1 = \frac{2144.95}{(53.3)(114+460)\,°R} = .0701 \text{ LBM/FT}^3$$

$$P_2 = (30.05 + .287 - .838)(.491)(144) = 2085.7 \text{ LBF/FT}^2$$

NOTE THAT THE STATIC AND ATMOSPHERIC PRESSURES CANCEL IN Δp.

$$\dot{m} = (1.012)(.0123)(0.0701) \times \sqrt{2\left(\frac{32.2}{.0701}\right)(2144.95 - 2085.7)}$$
$$= 0.204 \text{ LBM/SEC}$$

24 FROM EQN 4.120

$$\Delta p = \frac{\rho}{2g}\left(\frac{Q}{C_f A_o}\right)^2$$

$$A_o = \frac{\pi}{4}\left(\frac{.2}{12}\right)^2 = .000218 \text{ FT}^2$$

$$A_p = \frac{\pi}{4}\left(\frac{1}{12}\right)^2 = .00545 \text{ FT}^2$$

$$Q = vA = (1) \text{ FT/SEC} (.00545) \text{ FT}^2 = .00545 \text{ FT}^3/\text{SEC}$$

THEN

$$\Delta p = \frac{55}{(2)(32.2)}\left(\frac{.00545}{(.6)(.000218)}\right)^2 = 1482.7 \text{ PSF}$$

25 FROM EQUATIONS 4.120 AND 4.123,

$$\Delta p = \frac{\rho}{2g}\left(\frac{Q}{C_f A_o}\right)^2$$

ASSUME 70°F WATER. FROM PAGE 4-32,

$\nu = 1.059 \text{ EE-5}$

$$N_{Re} = \frac{Dv}{\nu} = \frac{(1)(2)}{1.059 \text{ EE-5}} = 1.89 \text{ EE5}$$

$$A_o = \frac{\pi}{4}(.2)^2 = .0314$$

$$A_p = \frac{\pi}{4}(1)^2 = .7854$$

$$\frac{A_o}{A_p} = \frac{.0314}{.7854} = .04$$

(MORE)

PROBLEM 25 CONTINUED

FROM FIGURE 4.23, $C_f \approx .60$

$Q = AV = (.7854)(2) = 1.571$

SO $\Delta p = \frac{62.4}{(2)(32.2)}\left(\frac{1.571}{(.6)(.0314)}\right)^2 = 6737.4$ PSF

26 THE HEIGHT IS A FUNCTION OF THE PRESSURE ONLY.

$h = \frac{p}{\rho} = \frac{(28)(144)}{56} = 72'$ FROM CENTER OF PIPE

27 $\rho_{AIR} = P/RT = \frac{(14.7)(144)}{(53.3)(460+70)} = .0749$ LBM/FT³

$\rho_{Hg} = (.491)\frac{\text{LBM}}{\text{IN}^3}(12)^3 = 848.4$ LBM/FT³

THE RISE IN FLUID WAS

$\left(\frac{2}{12}\right)$ FT $= .1667$ FT

FROM EQN 4.113

$v = \sqrt{\frac{(2)(32.2)(848.4 - .0749)(.1667)}{(.0749)}} = 348.7$ FPS

28 $A_A = \frac{\pi}{4}\left(\frac{24}{12}\right)^2 = 3.142$ FT²

$A_B = \frac{\pi}{4}\left(\frac{12}{12}\right)^2 = .7854$ FT²

$V_A = \frac{Q}{A} = \frac{8}{3.142} = 2.546$ FPS

$V_B = \frac{8}{.7854} = 10.19$ FPS

$P_A = (20)\text{ FT}(62.4)\frac{\text{LBM}}{\text{FT}^3} = 1248$ LBF/FT²

FROM EQN 4.55

$\Delta\left(\frac{P}{\rho}\right) = \Delta\left(\frac{v^2}{2g}\right)$

OR $P_B = 1248 - \left(\frac{(10.19)^2 - (2.546)^2}{2(32.2)}\right)62.4$

$= 1153.67$ PSF

FROM EQN 4.160 WITH $\phi = 0$

$F_x = (1248)(3.142) - (1153.67)(.7854) - \frac{(8)(62.4)}{32.2}(10.19 - 2.546)$

$= 2897$ LBF ON FLUID TO RIGHT

$F_y = 0$

29 $A = \frac{.75}{144} = .005208$

$Q = Av = (.005208)(250) = 1.302$ FT³/SEC

$\dot{w} = \rho Q = (.075)(1.302) = .09765$ LBM/SEC

$F = \dot{m}\,\Delta V_x = \frac{(.09765)\frac{\text{LBM}}{\text{SEC}}(250)\text{ FT/SEC}}{(32.2)\text{ FT/SEC}^2}$

$= .758$ LBF TO LEFT

30 $\dot{w} = (100)\text{ FT}^3/\text{SEC}\,(62.4)\frac{\text{LBM}}{\text{FT}^3} = 6240$ LBM/SEC

a) $F = \dot{m}\,\Delta V_x = \frac{(6240)(0-600)}{32.2} = 1.163$ EE5 LBF

b) $F = \dot{m}\,\Delta V_x = \frac{(6240)(-600-600)}{32.2} = 2.325$ EE5 LBF

31 FROM EQN 4.165

$D = \frac{C_D A \rho v^2}{2g}$

FOR STANDARD AIR

$P = 14.7$ PSIA

$T = 70°$F

$\rho = P/RT = \frac{(14.7)(144)}{(53.3)(460+70)°R} = .0749$ LBM/FT³

$A = \frac{\pi}{4}(1)^2 = .7858$

$D = \frac{(.2)(.7858)(.0749)(100)^2}{(2)(32.2)} = 1.83$ LBF

32 $\frac{(90)\text{ MPH}(5280)\text{ FT/MILE}}{(3600)\text{ SEC/HR}} = 132$ FT/SEC

ASSUME $\rho = .075$ LBM/SEC

FROM EQN 4.164

$5000 = \frac{(.5)(A)(.075)(132)^2}{(2)(32.2)}$

OR $A = 492.8$ FT²

33 SINCE $F \propto v^2$, THE FORCE WILL INCREASE BY A FACTOR OF 4 WHEN THE VELOCITY DOUBLES.

$F_2 = 4(800) = 3200$ LBF

34 FROM EQN 4.171

$N_{Re,M} = N_{Re,t}$

$\left(\frac{vD}{\nu}\right)_m = \left(\frac{vD}{\nu}\right)_t$ BUT $\nu_m = \nu_t$

$(V_M)\left(\frac{1}{10}D_t\right) = (60)\text{ MPH}(D_t)$

$V_M = 600$ MPH

35 FROM EQN 4.171

$N_{Re,m} = N_{Re,t}$

$\left(\frac{vL}{\nu}\right)_m = \left(\frac{vL}{\nu}\right)_t$ ASSUME $\nu_m = \nu_t$

$V_m(10) = (10)(600)$

OR $V_m = 600$ FPS

36

$A = 2d$

$r_H = \frac{2d}{2+2d} = \frac{d}{1+d}$

FROM EQN 5.18

$Q = \frac{1.49}{N} A (r_H)^{2/3} \sqrt{S}$

$3 = \left(\frac{1.49}{.012}\right)(2d)\left(\frac{d}{1+d}\right)^{2/3} \sqrt{.01}$

$\left(\frac{d^{5/2}}{1+d}\right)^{2/3} = .1208$

BY TRIAL AND ERROR, $d = .314$ FT

37 $A = \left(\frac{1}{2}\right)\frac{\pi}{4}\left(\frac{18}{12}\right)^2 = .8836$ FT^2

$r_H = \frac{\left(\frac{18}{12}\right)}{4} = .375$

FROM EQN 5.18

$Q = \frac{1.49}{.013}(.8836)(.375)^{2/3}\sqrt{.001}$

$= 1.665$ CFS

38 ASSUME FULL FLOWING PIPE. FROM EQN 5.19

$(100)\text{ gpm} \doteq (100)(.002228) = .2228$ FT^3/SEC

$D = (1.33)\left[\frac{(.2228)(.017)}{\sqrt{.02}}\right]^{3/8} = .343$

E-I-T HOMEWORK SOLUTIONS: THERMODYNAMICS

1 $F = ma$

$$(4)\ \text{LBF} = \frac{(10)\ \text{LBM}}{(32.2)\ \frac{\text{FT}}{\text{SEC}^2}}\ a$$

$$a = \frac{(4)(32.2)}{(10)}\ \frac{\text{LBF}\ \frac{\text{FT}}{\text{SEC}^2}}{\text{LBM}} = 12.88\ \frac{\text{FT}}{\text{SEC}^2}$$

$$\frac{(12.88)\ \text{FT}\ (60)^2\ \text{SEC}^2}{\text{SEC}^2\ \ \text{MIN}^2} = 46368\ \frac{\text{FT}}{\text{MIN}^2}$$

2

$$\frac{(.70)(140)\ \frac{\text{LBM}}{\text{FT}^3} + (.30)(62.4)\ \frac{\text{LBM}}{\text{FT}^3}}{32.2\ \frac{\text{LBM-FT}}{\text{LBF-SEC}^2}}$$

$$= 3.62\ \frac{\text{LBF-SEC}^2}{\text{FT}^4} = 3.62\ \frac{\text{SLUG}}{\text{FT}^3}$$

3 FROM PAGE 1-38, MOON GRAVITY IS 5.47 FT/SEC^2, FROM EQN 12.2,

$$F = \frac{ma}{g_c} = \frac{(10)(5.47)}{32.2} = 1.7\ \text{LBF}$$

4 ON MOON

$$\text{WEIGHT} = (10)\ \text{SLUGS}\ (5.47)\ \frac{\text{FT}}{\text{SEC}^2} = 54.7\ \text{LBF}$$

ON EARTH

$$\text{WEIGHT} = (10)\ \text{SLUGS}\ (32.2)\ \frac{\text{FT}}{\text{SEC}^2} = 322.0\ \text{LBF}$$

5 $F = ma$

$$= (11)\ \text{kg}\ (.8)\ \text{M/sec}^2 = 8.8\ \frac{\text{KG-M}}{\text{SEC}^2}$$

$$= 8.8\ N$$

6

$$F = (40)\ \text{SLUG}\ (8)\ \frac{\text{FT}}{\text{SEC}^2} = 320\ \text{LBF}$$

7 FROM PAGE 1-38,

$$K = 1.38\ \text{EE-23}\ \frac{\text{J}}{°\text{K}}$$

$$(1.38\ \text{EE-23})\ \frac{\text{J}}{°\text{K}}\ (1)\ \frac{\text{KG-M}^2}{\text{SEC}^2\text{-J}}\ (.06852)\ \frac{\text{SLUG}}{\text{KG}}\ (3.281)^2 \left(\frac{\text{FT}}{\text{M}}\right)^2 \left(\frac{5}{9}\right) \frac{°\text{K}}{°\text{R}}$$

$$\times\ (1)\ \frac{\text{LBF-SEC}^2}{\text{SLUG-FT}} = 5.66\ \text{EE-24}\ \frac{\text{FT-LBF}}{°\text{R-molecule}}$$

8 FROM PAGE 6-5,

a) $°\text{K} = 20°\text{C} + 273.15 = 293.15°\text{K}$

b) $°\text{F} = 32 + \frac{9}{5}(20) = 68°\text{F}$

c) $°\text{R} = 70 + 459.67 = 529.67°\text{R}$

9 CONSTANT VOLUME PROCESS (PAGE 6-24)

$$P_2 = \frac{P_1 T_2}{T_1} = \frac{(76)\ \text{CM}\ (100+273)°\text{K}}{(25+273)°\text{K}}$$

$$= 95.1\ \text{CM}$$

10 FROM EQUATION 6.37,

$$\frac{P_1 V_1}{T_1} = \frac{P_2 V_2}{T_2}$$

STP = 32°F = 492°R, 14.7 PSIA

$$T_2 = \frac{P_2 V_2 T_1}{P_1 V_1} = \frac{(15)\ \text{PSIA}\ (1.5)\ \text{FT}^3\ (492)°\text{R}}{(14.7)\ \text{PSIA}\ (1.2)\ \text{FT}^3}$$

$$= 167.55°\text{F}$$

11 FROM EQUATION 6.44,

$$P = \rho RT \quad \text{OR} \quad R = \frac{P}{\rho T}$$

$$R = \frac{(2)\ \text{ATM}\ (14.7)\frac{\text{LBF}}{\text{IN}^2\text{-ATM}}\ (144)\ \frac{\text{IN}^2}{\text{FT}^2}}{(.094)\ \frac{\text{LBM}}{\text{FT}^3}\ (100+460)°\text{R}}$$

$$= 80.43\ \text{FT}/°\text{R}$$

THEN, ALSO FROM EQN 6.44,

$$P = R\rho T = \frac{(80.43)\ \frac{\text{FT}}{°\text{R}}\ (.270)\ \frac{\text{LBM}}{\text{FT}^3}\ (250+460)°\text{R}}{(14.7)\ \frac{\text{LBF}}{\text{IN}^2\text{ATM}}\ (144)\ \frac{\text{IN}^2}{\text{FT}^2}}$$

$$= 7.28\ \text{ATM}$$

12 FROM EQN 6.37,

$$P_2 = \frac{P_1 V_1 T_2}{T_1 V_2} = \frac{(24+14.7)\ \text{PSIA}\ (1000)\ \text{IN}^3\ (460+35)}{(1020)\ \text{IN}^3\ (460+32)}$$

$$= 38.17\ \text{PSIA}$$

13 FROM EQN 6.42,

$$P = \frac{m R^* T}{V}$$

$$m = \frac{4.87\ \text{g}}{2\ \text{g/gmole}} = 2.435\ \text{gmoles}$$

$$R^* = .08206\ \frac{\text{ATM-LITER}}{\text{gmole-}°\text{K}} \quad \text{(TABLE 6.3)}$$

– MORE –

PROBLEM 13 CONTINUED

$T = (10 + 273) = 283\,°K$

$V = 6.8\ \ell$

$$P = \frac{(2.435)\ \text{gmoles}\ (.08206)\ \frac{\text{ATM-LITER}}{\text{gmole-}°\text{K}}\ (283)\ °\text{K}}{(6.8)\ \text{LITER}}$$

$= 8.31$ ATM

14 $V = \frac{4}{3}\pi r^3 = \left(\frac{4}{3}\right)(\pi)(30)^3 = 113{,}097$

THE LIFTING POWER IS THE WEIGHT OF THE DISPLACED AIR MINUS THE HELIUM WT. FROM EQUATION 6.42

$$M = \frac{PV}{RT}$$

FROM PAGE 6-12, $R_{AIR} = 53.3$

$R_{helium} = 386.3$

LIFTING POWER =

$$\frac{(14.7)\ \text{PSIA}\ (144)\ \frac{\text{IN}^2}{\text{FT}^2}\ (113{,}097)\ \text{FT}^3}{(32 + 460)\ °\text{R}}\left[\frac{1}{(53.3)\ \text{FT}/°\text{R}} - \frac{1}{(386.3)}\right]$$

$= 7869.7$ LBF

15 END AREA $= \frac{\pi}{4}d^2 = \frac{\pi}{4}(4)^2 = 12.56\ \text{IN}^2$

PRESSURE DIFFERENCE $= \left(\frac{2}{3}\right)(14.7)$ PSIA

$\Delta\text{FORCE} = \Delta PA = \left(\frac{2}{3}\right)(14.7)\ \text{PSIA}\ (12.56)\ \text{IN}^2$

$= 123.1$ LBF

16 FROM EQN 6.42, THE ORIGINAL WEIGHT OF AIR IN THE TANK IS

$$M = \frac{PV}{RT} = \frac{(120 + 14.7)\ \text{PSI}\ (144)\ \text{IN}^2/\text{FT}^2\ (3)\ \text{FT}^3}{(53.3)\ \text{FT}/°\text{R}\ (460° + 80°)\ °\text{R}} = 2.022\ \text{LBM}$$

THE FINAL WEIGHT OF THE AIR IN THE TANK IS

$$\frac{(28 + 14.7)(144)(3)}{(53.3)(540)} = .641\ \text{LBM}$$

THE AIR USED IN THE TIRES IS

$2.022 - .641 = 1.381$ LBM

THE WEIGHT OF AIR PER TIRE IS

$$\frac{(28 + 14.7)(144)(1.2)}{(53.3)(540)} = .256\ \text{LBM}$$

$$\#\ \text{TIRES} = \frac{1.381}{.256} = 5.39$$

SAY 5 COMPLETE TIRES

17 FOR COPPER $c = .091$ BTU/LBM-°F

FROM EQUATION 6.18

$Q = cm\Delta T$

$= (.091)\ \frac{\text{BTU}}{\text{LBM-}°\text{F}}\ (20)\ \text{LBM}\ (500 - 30)\ °\text{F}$

$= 855.4$ BTU

18 FOR ALUMINUM, $c = .208$ cal/g-°C

FROM EQN 6.18

$Q = cm\Delta T$

$= (.208)\ \frac{\text{cal}}{\text{g-}°\text{C}}\ (20)\ \text{KG}\ (1000)\ \text{g/kg}\ (600 - 80)\ °\text{C}$

$= 2.163$ EE6 cal

19 30 kg = 3EE4 g

TO RAISE TEMPERATURE OF WATER FROM 20°C TO 100°C

$Q = (1)\ \frac{\text{cal}}{\text{g-}°\text{C}}\ (3\text{EE}4)\ \text{g}\ (100 - 20)\ °\text{C}$

$= 2.4$ EE6 cal

TO VAPORIZE THE WATER (TABLE 6.2)

$h_{fg} = (3\text{EE}4)\ \text{g}\ (539.55)\ \frac{\text{cal}}{\text{g}}$

$= 1.62$ EE7 cal

TOTAL = 2.4EE6 + 1.62 EE7

= 1.86 EE7 cal

20 FROM TABLE 6.4

AIR $c_p = .241$ BTU/LBM-°F

$c_v = .1724$ BTU/LBM-°F

a) $Q = cm\Delta T$

$= (.1724)\ \frac{\text{BTU}}{\text{LBM-}°\text{F}}\ (4)\ \text{LBM}\ (180 - 70)\ °\text{F}$

$= 75.86$ BTU

b) $Q = (.241)(4)(180 - 70)$

$= 106.04$ BTU

21

1 GALLON = .1337 FT³ (ALSO, SEE P. 1-36)

WATER WEIGHT = (.1337)(62.4)

= 8.343 LBM

USING EQNS 6.106 AND 6.107

$Q_{\text{LOST BY IRON}} = Q_{\text{GAINED BY WATER}}$

$(.1)(2)(200 - T_2) = (1)(8.343)(T_2 - 40)$

$40 - .2T_2 = 8.343\,T_2 - 333.72$

$T_2 = 43.75\ °\text{F}$

22 ICE $c = .5$ (AVERAGE)

STEAM $c = .5$ (AVERAGE)

THE HEAT REQUIRED TO RAISE THE ICE FROM 0°F TO 32°F IS

$$Q = (.5)\frac{\text{BTU}}{\text{LBM-°F}}(20)\text{LBM}(32-0) = \underline{320 \text{ BTU}}$$

THE HEAT REQUIRED TO MELT THE ICE, FROM TABLE 6.2, IS

$$Q = (20)\text{LBM}(143.4)\frac{\text{BTU}}{\text{LBM}} = \underline{2868 \text{ BTU}}$$

THE HEAT TO RAISE THE WATER FROM 32°F TO 212°F IS

$$Q = (1)(20)(212-32) = \underline{3600 \text{ BTU}}$$

THE HEAT TO VAPORIZE THE WATER (FROM TABLE 6.2 AGAIN) IS

$$Q = (20)(970.3) = \underline{19406} \text{ BTU}$$

THE HEAT TO RAISE STEAM FROM 212°F TO 213°F IS

$$Q = (.5)(20)(213-212) = \underline{10 \text{ BTU}}$$

TOTAL HEAT REQUIRED IS

$$320 + 2868 + 3600 + 19406 + 10 = 26204 \text{ BTU}$$

THE TIME IS

$$t = \frac{26204 \text{ BTU}}{160 \text{ BTU/SEC}} = 163.8 \text{ SEC}$$

23

$\dot{m}_1, h_1$; $\dot{m}_2 h_2$; $(\dot{m}_1 + \dot{m}_2)h_3$

FROM PAGE 6-30, THE ENTHALPY OF THE INCOMING WATER IS

$$h_1 = 48.02 \text{ BTU/LBM}$$

FROM PAGE 6-31, THE ENTHALPY OF THE INCOMING STEAM IS

$$h_2 = 298.40 + .8(888.8) = 1009.44 \text{ BTU/LBM}$$

FROM PAGE 6-30, THE MIXTURE ENTHALPY IS

$$h_3 = 147.92 \text{ BTU/LBM}$$

THEN,

$$\left.\begin{aligned}\dot{m}_1 + \dot{m}_2 &= 4000 \text{ LBM/HR}\\ 48.02\dot{m}_1 + 1009.44\dot{m}_2 &= 4000(147.92)\end{aligned}\right\}$$

SOLVING SIMULTANEOUSLY,

$$\dot{m}_1 = 3584.4 \text{ LBM/HR}$$

$$\dot{m}_2 = 415.6 \text{ LBM/HR}$$

24 THIS MUST BE SOLVED BY EQN. 6.106. LET X BE THE VOLUME FLOW IN GPM.

$$Q_{\text{LOST BY A}} = Q_{\text{GAINED BY B}}$$

$$(X_A)\text{gpm}(10)\frac{\text{BTU}}{\text{gal-°F}}(140-80)\text{°F}$$

$$= (X_B)(8.33)(80-65)$$

$$\left.\begin{aligned}\text{OR} \quad 600X_A &= 124.95X_B\\ \text{ALSO,} \quad X_A + X_B &= 50\end{aligned}\right\}$$

SOLVING SIMULTANEOUSLY,

$$X_A = 8.62 \text{ gpm}$$

$$X_B = 41.38 \text{ gpm}$$

FROM AN ENERGY BALANCE,

$$(8.62)\text{gpm}(10)\frac{\text{BTU}}{\text{gal-°F}}(140)\text{°F} + (41.38)(8.33)(65)$$

$$= (50)(c)(80)$$

$$c = 8.62 \text{ BTU/gal-°F}$$

25 FROM EQUATION 6.46,

$$c_v = c_p - \frac{R}{J}$$

$$= .245\frac{\text{BTU}}{\text{LBM-°R}} - \frac{533\frac{\text{FT-LBF}}{\text{LBM-°R}}}{778\frac{\text{FT-LBF}}{\text{BTU}}}$$

$$= .1765\frac{\text{BTU}}{\text{LBM-°R}}$$

$$\text{THEN,} \quad Q = (10)\text{LBM}(.1765)\frac{\text{BTU}}{\text{LBM-°R}}(360-90)\text{°F}$$

$$= 476.6 \text{ BTU}$$

26 FOR CONSTANT VOLUME PROCESSES,

$$Q = Nc_v\Delta T$$

$$(200)\text{BTU} = (.5)\text{moles}(c_v)(100)\text{°F}$$

$$c_v = (4)\text{ BTU/mole-°F}$$

FROM EQN 6.47,

$$c_p = \frac{R^*}{J} + c_v$$

$$= \left(\frac{1545}{778}\right)\frac{\text{BTU}}{\text{pmole-°R}} + (4)\frac{\text{BTU}}{\text{pmole-°R}}$$

$$= (5.99)\frac{\text{BTU}}{\text{pmole-°R}}$$

THE QUANTITY $\frac{1545}{778}$ IS EQUAL TO 1.986, AS LISTED IN TABLE 6.3.

27 FROM EQN 6.98

$$\Delta U = Q - (W/J)$$

$$= (10)\,\text{BTU} - \frac{(1500)\ \text{FT-LBF}}{(778)\ \frac{\text{FT-LBF}}{\text{BTU}}}$$

$$= 8.07\ \text{BTU}\ \{\text{INCREASE}\}$$

28 $Q = \Delta U + W$ BUT $\Delta U = 0$

AND $W = \Delta E_K = \frac{1}{2}mv^2$

$$(500)\ \text{cal}\ (4.186)\frac{\text{Joules}}{\text{cal}} = (5)(5)g\left(\frac{1}{1000}\right)\frac{\text{KG}}{g}v^2$$

$$v^2 = (837200)\frac{\text{Joules}}{\text{kg}} = 837200\ \frac{\text{kg-m}^2}{\text{kg-sec}^2}$$

$$v = 914.99\ \text{m/sec}$$

29 $Q = \Delta U + W$ AND $W = \Delta$ POTENTIAL ENERGY

$$0 = wC_p\Delta T + \frac{wh}{J}$$

ON A 1-POUND BASIS, THE CHANGE IN POTENTIAL ENERGY IS

$$\frac{(1)\ \text{LBM}\ (-200)\ \text{FT}}{778\ \frac{\text{FT-LBF}}{\text{BTU}}} = -.257\ \text{BTU}\ (\text{decrease})$$

THE CHANGE IN INTERNAL ENERGY IS

$$(.257)\ \text{BTU} = (1)\ \text{LBM}\ (1)\frac{\text{BTU}}{\text{LBM-}^\circ\text{F}}\ \Delta T$$

$$\Delta T = .257^\circ\text{F}$$

30 $Q = \Delta U + W$

$$Q = \left(\frac{1}{10}\right)\text{HP}\,(33000)\frac{\text{FT-LBF}}{\text{MIN-HP}}\left(\frac{1}{778}\right)\frac{\text{BTU}}{\text{FT-LBF}}(15)\ \text{MIN}$$

$$= 63.62\ \text{BTU}$$

$$(63.62)\ \text{BTU}\ (252)\frac{\text{cal}}{\text{BTU}} = 16032.2\ \text{cal}$$

$W = 0$

$c = .2\ \text{cal/g-}^\circ\text{C}$ FOR GLASS

$$\Delta U = (200)\,g\,(.2)\frac{\text{cal}}{g\text{-}^\circ\text{C}}(T_2 - 20)$$

$$+ (2000)\,g\,(1)\frac{\text{cal}}{g\text{-}^\circ\text{C}}(T_2 - 20)$$

$$= 2040(T_2 - 20)\ \text{cal}$$

SO $16032.2 = 2040(T_2 - 20)$ OR $T_2 = 27.86^\circ\text{C}$

31 $\text{ENERGY} = (.5)(1)\ \text{LBM}\left(\frac{1}{32.2}\right)\frac{\text{SLUG}}{\text{LBM}}\left[(30)^2 - (10)^2\right]\ \text{FT}^2/\text{sec}^2$

$$+ (1)\ \text{LBM}\ (0 - 40)\ \text{FT}$$

$$= -27.58\ \text{FT-LBF/LBM}$$

FROM PAGE 1-36

$$(-27.58)\frac{\text{FT-LBF}}{\text{LBM}}(3.766\ \text{EE-7})\frac{\text{KW-HRS}}{\text{FT-LBF}}$$

$$= -1.039\ \text{EE-5}\ \frac{\text{KW-HRS}}{\text{LBM}}$$

32

a) FROM PAGE 6-30 $h = 48.02$ BTU/LBM

b) " " 6-32 $h = 1533$ "

c) " " 6-30 $h = 1179.7$ "

d) " " 6-31 $h = 1164.1$ "

e) " " 6-38 $h = 662.1$ "

f) THIS REQUIRES DOUBLE INTERPOLATION FROM PAGE 6-41

AT 40 PSIA, 170°F

$$h \approx 99.83 + \frac{2}{5}(107.88 - 99.83) = 103.05$$

AT 50 PSIA, 170°F

$$h \approx 99.54 + \frac{2}{5}(107.62 - 99.54) = 102.77$$

SO, AT 45°F, 170°F

$$h = \frac{1}{2}(103.05 + 102.77) = 102.91\ \frac{\text{BTU}}{\text{LBM}}$$

g) $400^\circ\text{F} = 860^\circ\text{R}$

FROM PAGE 6-36, $h = 206.46$ BTU/LBM

33 FROM PAGE 6-30

$$h_1 = 48.02\ \text{BTU/LBM}$$

$$h_2 = 48.02 + \frac{240\ \text{BTU}}{4\ \text{LBM}} = 108.02\ \text{BTU/LBM}$$

34

a) FROM PAGE 6-31 $S = .3680$ BTU/LBM-°F

b) " " 6-32 $S = 2.0456$

c) IF EXPANSION IS ISENTROPIC, $\Delta S = 0$

SO S AT 100 PSIA AND 800°F

$= 1.8443$

d) USE THE MOLLIER DIAGRAM IN A PROCEDURE DESCRIBED ON PAGES 7-4 AND 7-5 TO FIND h_2 ASSUMING ISENTROPIC EXPANSION.

$$h_2 = 1131\ \text{BTU/LBM}$$

BUT, THE PROCESS IS NOT ISENTROPIC, AND THE ACTUAL ENTHALPY IS

$$h_2' = 1428.9 - .8(1428.9 - 1131) = 1190.7$$

FROM THE MOLLIER DIAGRAM {PAGE 6-35} WITH $P = 5$ PSIA AND $h = 1190.7$,

$$S \approx 1.93\ \text{BTU/LBM-}^\circ\text{F}$$

35 THIS IS SIMILAR TO EXAMPLE 6.17

$h_1 = 1530.8$ BTU/LBM

$S_1 = 1.9193$ BTU/LBM-°R

$T_0 = 60 + 460 = 520$ R

h_2 {FOR 60°F WATER} $= 28.06$ BTU/LBM

$S_2 = .0555$ BTU/LBM-°R

{MORE}

PROBLEM 35 CONTINUED

FROM EQN 6.11

AVAILABILITY = (1530.8 − 28.06) BTU/LBM − 520°R (1.9193 − .0555) BTU/LBM-°R = 533.56 BTU/LBM

36 THE 900°F DATA IS IRRELEVANT.

$T_0 = 80 + 460 = 540°R$

ASSUME FEEDWATER IS AT 80°F.

$h_{WATER} = 48.02$ BTU/LBM

$S_{WATER} = .0932$ BTU/LBM-°F

THE STEAM ENTHALPY IS

$h_{STEAM} = 1201.0$ BTU/LBM

$S_{STEAM} = 1.5272$ BTU/LBM-°R

THE INCREASE IS $(1201 - 48.02) = 1152.98$

THE AVAILABILITY IS

$(1201.0 - 48.02) - 540(1.5272 - .0932) = 378.62$ BTU/LBM

THE MAXIMUM WORK PER INPUT BTU IS

$$\frac{378.62 \frac{BTU}{LBM}}{1152.98 \frac{BTU}{LBM}} = .328 \frac{BTU}{BTU}$$

37 $T_1 = 300 + 460 = 760°R$

$T_0 = 80 + 460 = 540°R$

FROM PAGE 6-36

$h_1 = 182.08$ BTU/LBM

$\phi_1 = .68312$ BTU/LBM-°R

$P_1 = 40$ PSIA

$h_2 = 129.06$

$\phi_2 = .60078$

$P_2 = 14.7$

$S_2 - S_1 = \phi_2 - \phi_1 - \left(\frac{R}{J}\right) \ln\left(\frac{P_2}{P_1}\right)$ {EQN. 6.29}

$$= .60078 - .68312 - \left(\frac{53.3 \text{ FT/°R}}{778 \frac{\text{FT-LBF}}{\text{BTU}}}\right) \ln\left(\frac{14.7}{40}\right)$$

$= -.01376$

OR $S_1 - S_2 = .01376$

FROM EQN. 6.11 AVAILABILITY IS

$(182.08 - 129.06) - 540(.01376) = 45.59$ BTU/LBM

38 FROM PAGE 6-25

$Q = P_1 V_1 \ln\left(\frac{V_2}{V_1}\right)$

BUT $PV = wRT$

$Q = wRT \ln\left(\frac{V_2}{V_1}\right)$

$R = 55.2$ {FROM PAGE 6-12}

$$Q = \frac{(3) \text{ LBM} (55.2) \frac{\text{FT-LBF}}{\text{LBM-°R}} (760)°R \ln\left(\frac{22.5}{40}\right)}{778 \frac{\text{FT-LBF}}{\text{BTU}}}$$

$= -93.08$ BTU {OUT OF SYSTEM}

39 FROM PAGE 6-25 FOR A CLOSED POLYTROPIC PROCESS,

$$W = \frac{P_1 V_1 - P_2 V_2}{N - 1}$$

$$P_2 = P_1 \left(\frac{V_1}{V_2}\right)^N = (200) \text{ PSI} \left(\frac{.02}{.028}\right)^{-1.2} = 299.5 \text{ PSIA}$$

$$W = \frac{(144) \frac{\text{IN}^2}{\text{FT}^2} \left[(20) \text{ PSI} (.02) \text{FT}^3 - (299.5)(.028)\right]}{-1.2 - 1.0}$$

$= 287.08$ FT-LBF

40 $P_2 = 2P_1$

FROM PAGE 6-25

$W = P_1 V_1 \ln\left(\frac{V_2}{V_1}\right)$

BUT $\frac{V_2}{V_1} = \frac{P_1}{P_2}$ {BOYLE'S LAW}

AND $PV = wRT$

SO $W = wRT_1 \left(\frac{P_1}{P_2}\right)$

$$= \frac{(4) \text{ LBM} (53.3) \frac{\text{FT-LBF}}{\text{LBM-°R}} (700)°R \ln\left(\frac{1}{2}\right)}{778 \frac{\text{FT-LBF}}{\text{BTU}}}$$

$= -132.96$ BTU

41 REVERSIBLE ADIABATIC = ISENTROPIC

$Q = 0$

42 ADIABATIC COMPRESSION

$V_1 = 2$ $V_2 = 1$ $T_1 = 20 + 460 = 480°R$

$T_2 = 480\left(\frac{2}{1}\right)^{1.4-1} = 633.4$ °R

$W = C_v \Delta T = .1724 (480 - 633.4) = -26.45 \frac{BTU}{LBM}$

$= -20578 \frac{\text{FT-LBF}}{\text{LBM}}$

PROBLEM 42 CONTINUED

$\Delta U = -W = 20578 \ \frac{FT\text{-}LBF}{LBM}$

$Q = 0$ SINCE ADIABATIC

$$P_1 = \frac{wRT_1}{V_1} = \frac{(1)(53.3)(480)}{2} = 12792 \text{ PSF}$$

$$P_2 = P_1\left(\frac{633.4}{480}\right)^{\frac{1.4}{1.4-1}} = 33765 \text{ PSF}$$

ISOBARIC HEAT ADDITION

$V_1 = 1 \quad V_2 = 2 \quad T_1 = 633.4$

$P_1 = P_2 = 33765$ PSF

$T_2 = 633.4\left(\frac{2}{1}\right) = 1266.8\ °R$

$W = 33765(2-1) = 33765 \ \frac{FT\text{-}LBF}{LBM}$

$$\Delta U = \frac{P(V_2 - V_1)}{K-1} = \frac{(33765)(2-1)}{1.4-1} = 84413 \text{ FT-LBF}$$

$Q = C_p \Delta T = .241(1266.8 - 633.4) = 152.65$ BTU

$= 118{,}762 \ \frac{FT\text{-}LBF}{LBM}$

CONSTANT VOLUME

$P_1 = 33765 \quad P_2 = 12792 \quad V_1 = V_2 = 2$

$T_1 = 1266.8\ °R \quad T_2 = 480\ °R$

$W = 0$

$\Delta U = C_v(\Delta T) = .1724(480 - 1266.8) = -135.64$ BTU

$= -105{,}528 \ \frac{FT\text{-}LBF}{LBM}$

$Q = \Delta U = -105{,}528 \ \frac{FT\text{-}LBF}{LBM}$

43 $pV = wRT$ OR

$$v_1 = \frac{(1)\text{LBM}(53.3)\frac{FT\text{-}LBF}{LBM\text{-}°R}(460+70)\ °R}{(14.7)\text{ PSI }(144)\frac{IN^2}{FT^2}} = 13.35 \ FT^3/LBM$$

$v_2 = \frac{1}{10}(13.35) = 1.335 \ \frac{FT^3}{LBM}$

INTERPOLATING FROM PAGE 6-36 FOR 530° AIR,

$h_1 = 126.66$ BTU/LBM

$v_{r_1} = 151.22$

THEN $v_{r2} = \left(\frac{1}{10}\right) v_{r_1} = 15.122$

FINDING $v_r = 15.122$ IN THE TABLE,

$T_2 = 1292\ °R$ {IF ISENTROPIC}

$h_2 = 314.88$ {IF ISENTROPIC}

BUT $\eta_{ISEN} = .90$, SO

$$h_2' = 126.66 + \frac{314.88 - 126.66}{.9} = 335.79 \ \frac{BTU}{LBM}$$

SO $W_{COMPRESSION} = 335.79 - 126.66 = 209.13$ BTU/LBM

SEARCHING THE AIR TABLE FOR $h = 335.79$ LEADS TO $T_2' = 1372.5\ °R$

44 FROM PAGE 6-36 FOR $T_1 = (460+60) = 520\ °R$

$h_1 = 124.27$ BTU/LBM

$p_{r_1} = 1.2147$

$\phi_1 = .59173$ BTU/LBM-°R

FOR $T_2' = (460-100) = 360\ °R$

$h_2' = 85.97$

$\phi_2' = .50369$

SINCE $P_2 = 14.7$, P_{r_2} WOULD BE {IF ISENTROPIC}

$$P_{r_2} = \frac{14.7}{500}(1.2147) = .03571$$

BY USING AN AIR TABLE THAT GOES BELOW 360 °R,

T_2 {FOR $P_r = .03571$} $= 189\ °R$

$h_2 = 45.04$

$W = 85.97 - 124.27 = -38.3$ BTU/LBM

$$\Delta S = .50369 - .59173 - \left(\frac{53.3}{778}\right)\ln\left(\frac{14.7}{500}\right)$$

$= .1536 \ \frac{BTU}{LBM\text{-}°R}$

$$\eta_{ISEN} = \frac{124.27 - 85.97}{124.27 - 45.04} = .483$$

OR 48.3 %

45 FROM PAGE 6-26 FOR A THROTTLING PROCESS,

$$S_2 - S_1 = MR^* \ln\left(\frac{P_1}{P_2}\right)$$

$$= (3)\frac{\text{PMOLES}}{\text{SEC}}(1.986)\frac{\text{BTU}}{\text{PMOLE-}°R}\ln\left(\frac{9.2}{6.4}\right)$$

$$= 2.162 \ \frac{BTU}{SEC\text{-}°R}$$

46 $\Delta T = 0$ FOR THROTTLED IDEAL GASES

47 ASSUME $g_{local} = 32.2$ FT/SEC².

THE STEADY FLOW ENERGY EQUATION {SFEE} ON A ONE-POUND BASIS IS GIVEN BY EQN 6.104

$$q - \frac{W}{J} = h_2 - h_1 + \frac{v_2^2 - v_1^2}{2gJ} + \frac{z_2 - z_1}{J}$$

BUT $q = 0$

PROBLEM 47 CONTINUED

$T_1 = 1340 + 460 = 1800°R$

$h_1 = 449.71$ BTU/LBM

$T_2 = 820 + 460 = 1280°R$

$h_2 = 311.79$ BTU/LBM

$z_2 = z_1 = 0$

$$\text{SO } W = (-778)\ \frac{\text{FT-LBF}}{\text{BTU}}\left[(311.79)-(449.71)\ \frac{\text{BTU}}{\text{LBM}} + \frac{(1000)^2\ \frac{\text{FT}^2}{\text{SEC}^2} - (540)^2\ \frac{\text{FT}^2}{\text{SEC}^2}}{(2)(32.2)\frac{\text{FT}}{\text{SEC}^2}(778)\ \frac{\text{FT-LBF}}{\text{BTU}}}\right]$$

$$= 96{,}301.8\ \frac{\text{FT-LBF}}{\text{LBM AIR}} = 123.78\ \frac{\text{BTU}}{\text{LBM}}$$

48 $\dot{M} = 100{,}000$ LBM/HR $= 27.778$ LBM/SEC

$W = 10{,}000$ HP $= 7069.4\ \frac{\text{BTU}}{\text{SEC}}$

$h_1 = 1400$ BTU/LBM

$V_1 = 0 \quad V_2 = 500$ FPS

$Q = 0$

USING THE SFEE {ECQN 6.104}

$$-W = \dot{M}\left(h_2 - h_1 + \frac{V_2^2 - V_1^2}{2Jg}\right)$$

$$(-7069.4) = (27.778)\ \frac{\text{LBM}}{\text{SEC}}\left[h_2 - 1400 + \frac{(500)^2}{(2)(32.2)(778)}\right]$$

$$= 27.778(h_2 - 1395.0)$$

$$h_2 = 1140.5 \text{ BTU/LBM}$$

49 $\dot{M} = 275$ LBM/MIN $= 4.583$ LBM/SEC

$W = 1000$ HP $= 706.94$ BTU/SEC

USING THE SFEE {ECQN 6.104}

$$Q = (706.94) + (4.583)\left[(1012.5) - (1217.6) + \frac{(400)^2 - (70)^2}{(2)(32.2)(778)} + \frac{(1.8)-(10.2)}{(778)}\right]$$

$$= -218.9\ \frac{\text{BTU}}{\text{SEC}}$$

$$= -7.88 \text{ EE5 BTU/HR} \quad \{\text{LOSS}\}$$

50 USING THE SFEE {ECQN 6.104}

$$-50 - W = 5\left[(1020) - (1000) + \frac{(50)^2 - (100)^2}{(2)(32.2)(778)} + \frac{(0 - 100)}{(778)}\right]$$

$$-50 - W = 98.6$$

$$W = -148.6 \text{ BTU/SEC} \quad \{\text{INPUT}\}$$

$$= \frac{(148.6)\ \frac{\text{BTU}}{\text{SEC}}\ (778)\ \frac{\text{FT-LBF}}{\text{BTU}}}{550\ \frac{\text{FT-LBF}}{\text{HP-SEC}}}$$

$$= 210.2 \text{ HP}$$

51 FROM ECQN 6.54

$$V_{RMS} = \sqrt{\frac{3KT}{M}}$$

$$T = \left(\tfrac{5}{9}\right)(70 + 460) = 294.4\ °K$$

THE MOLECULAR WEIGHT OF OXYGEN IS 32 AMU. FROM PAGE 19-16, 1 AMU IS 1.66053 EE-27 KG

$$V_{RMS} = \sqrt{\frac{(3)(1.3803 \text{ EE-23})\frac{\text{KG-M}^2}{\text{SEC}^2\text{-}°\text{K}}(294.4)°\text{K}}{(32)\text{ AMU }(1.66053 \text{ EE-27})\text{ KG/AMU}}}$$

$$= 479 \text{ M/SEC}$$

52 FROM EQUATION 6.58

$$E_K = \left(\tfrac{3}{2}\right)(1.3803 \text{ EE-23})\ \frac{\text{J}}{°\text{K}}\left(\tfrac{5}{9}\right)(0 + 460)°\text{K}$$

$$= 5.29 \text{ EE-21 J/molecule}$$

53 FROM ECQN 6.53

$$V_M = 2\sqrt{\frac{2KT}{\pi M}}$$

$$M = (2)(14.007) = 28.014 \text{ AMU}$$

$$V_M = 2\sqrt{\frac{(2)(1.380 \text{ EE-23})\frac{\text{KG-M}^2}{\text{SEC}^2\text{-}°\text{K}}(20 + 273)°\text{K}}{(\pi)(28.014)\text{ AMU }(1.66053 \text{ EE-27})\text{ KG/AMU}}}$$

$$= 470.5 \text{ M/SEC}$$

54 USE TABLE 6.4 FOR THE GAS DATA

a) ASSUME SOME VOLUME – SAY 100 FT^3, CALCULATE THE WEIGHT OF EACH GAS AT ANY PRESSURE AND TEMPERATURE.

{MORE}

PROBLEM 54 CONTINUED

PV = WRT, so W = PV/RT

$W_{CO} = \left(\frac{P}{T}\right)\left(\frac{30}{55.2}\right) = .543\left(\frac{P}{T}\right)$ LBM

$W_{CO_2} = \left(\frac{P}{T}\right)\left(\frac{15}{35.1}\right) = .427\left(\frac{P}{T}\right)$ LBM

$W_{H_2} = \left(\frac{P}{T}\right)\left(\frac{55}{766.8}\right) = .0717\left(\frac{P}{T}\right)$ LBM

$$G_{CO} = \frac{.543\left(\frac{P}{T}\right)}{.543\left(\frac{P}{T}\right) + .427\left(\frac{P}{T}\right) + .0717\left(\frac{P}{T}\right)}$$

$$= \frac{.543}{1.0417} = .521$$

$$G_{CO_2} = \frac{.427}{1.0417} = .410$$

$$G_{H_2} = \frac{.0717}{1.0417} = .069$$

b) FROM TABLE 6.5, MOLECULAR WEIGHT IS VOLUMETRICALLY WEIGHTED

$$M = \frac{30(28.0) + 15(44.0) + 55(2.0)}{100} = 16.1$$

c) FROM TABLE 6.5, THE GAS CONSTANT IS GRAVIMETRICALLY WEIGHTED

$$R = \frac{.521(55.2) + .410(35.1) + .069(766.8)}{1.00}$$

$$= 96.06 \ \frac{\text{FT-LBF}}{\text{LBM-}^\circ\text{R}}$$

d) FROM TABLE 6.5, SPECIFIC HEATS ARE GRAVIMETRICALLY WEIGHTED

$$C_p = \frac{.521(.243) + .410(.205) + .069(3.42)}{1.00}$$

$$= .4467 \ \text{BTU/LBM-}^\circ\text{R}$$

$$C_v = \frac{.521(.1721) + .410(.1599) + .069(2.435)}{1.00}$$

$$= .3232 \ \text{BTU/LBM-}^\circ\text{R}$$

55 ASSUME SOME WEIGHT OF MIXTURE, SAY 100 POUNDS. THEN

$$N = \frac{\text{WEIGHT OF GAS}}{\text{MOLECULAR WEIGHT}}$$

$$N_{He} = \frac{20}{4.0} = 5$$

$$N_{AIR} = \frac{40}{29.0} = 1.379$$

$$N_{CO_2} = \frac{40}{44.0} = .9091$$

THEN, FROM EQN 6.69

$$B_{He} = \frac{5}{5 + 1.379 + .9091} = \frac{5}{7.2881} = .686$$

$$B_{AIR} = \frac{1.379}{7.2881} = .1892$$

$$B_{CO_2} = \frac{.9091}{7.2881} = .1247$$

b) $M = .686(4.0) + .1892(29.0) + .1247(44.0)$

$= 13.718$

c) $R = .20(386.3) + .40(53.3) + .40(35.1)$

$= 112.6 \ \frac{\text{FT-LBF}}{\text{LBM-}^\circ\text{R}}$

56 FROM EQN 6.42

$$W = p\left(\frac{V}{RT}\right)$$

$$W_{N_2} = p\left(\frac{(.6)(5)}{(55.2)(40+460)}\right) = (1.087 \text{ EE-4})\,p \text{ LBM}$$

$$W_{CO_2} = p\left(\frac{(.10)(5)}{(35.1)(500)}\right) = (2.849 \text{ EE-5})\,p \text{ LBM}$$

$$W_{H_2} = p\left(\frac{(.30)(5)}{(766.8)(500)}\right) = (3.912 \text{ EE-6})\,p \text{ LBM}$$

a) $W_T = p(1.087 \text{ EE-4} + 2.849 \text{ EE-5} + 3.912 \text{ EE-6})$

$= (1.411 \text{ EE-4})\,p$

$$G_{N_2} = \frac{(1.087 \text{ EE-4})\,p}{(1.411 \text{ EE-4})\,p} = .7704$$

$$G_{CO_2} = \frac{2.849 \text{ EE-5}}{1.411 \text{ EE-4}} = .2019$$

$$G_{H_2} = \frac{3.912 \text{ EE-6}}{1.411 \text{ EE-4}} = .0277$$

$C_p = .7704(.247) + .2019(.205) + .0277(3.42)$

$= .3264 \ \text{BTU/LBM-}^\circ\text{R}$

c) $Q = mC_p\Delta T = (1.411 \text{ EE-4})\,p(.3264)(250 - 40)$

$= (9.67 \text{ EE-3})\,p \ \text{BTU}$

57 a) T_{sat} FOR 80 PSIA STEAM = 312.03°F

$T = 312.03 + 388 = 700.03$°F

FROM PAGE 6-32

$h = 1379.9$ BTU/LBM

b) $h = h_f + x h_{fg}$. FROM PAGE 6-31

$= 298.4 + .45(888.8) = 698.36$ BTU/LBM

PROBLEM 57 CONTINUED

c) FROM PAGE 6-34

$h = 269.59 + 1.75 = 271.34$ BTU/LBM

d) FROM PAGE 6-32

$h = \frac{1}{2}(1431.3 + 1430.5) = 1430.9$ BTU/LBM

58 $x = \frac{h - h_f}{h_{fg}} = \frac{600 - 424}{780.5} = .225$

59 FROM PAGE 6-32

$v = 7.797$ FT³/LBM

60 USING THE MOLLIER DIAGRAM (PAGE 6-35) FOLLOWING A VERTICAL LINE DOWN,

≈ 5 PSIA

61 FROM PAGE 6-35 FOLLOWING A HORIZONTAL LINE TO THE RIGHT

≈ 2.0 BTU/LBM-°R

A FROM THE PSYCHROMETRIC CHART

a) 68°F

b) 48%

c) 86 GRAINS/LBM OR .0123 LBM/LBM

B SINCE $p = 10$ PSIA, WE CANNOT USE A PSYCHROMETRIC CHART. THE DEWPOINT TEMPERATURE IS ALSO THE SATURATION TEMPERATURE

AT $T_{SAT} = 50°$, $P_{SAT} = .1781$ PSIA

SO $P_{AIR} = 10 - .17811 = 9.822$ PSIA

a) $\omega = .622\left(\frac{P_{SAT}}{P_{AIR}}\right) = .622\left(\frac{.17811}{9.822}\right)$

$= .0113$ LBM/LBM

b) $\phi = \frac{P_{SAT}}{P_{SAT}|_{70°F}} = \frac{.17811}{.3631} = .4905$

c) $v = \frac{RT}{P} = \frac{(53.3)(530)}{9.822} = 19.98$ FT³/LBM

C FROM THE PSYCHROMETRIC CHART,

$h_1 = 41.4$ BTU/LBM

$h_2 = 20.8$ BTU/LBM

$\omega_1 = 118$ GRAINS/LBM

$= \frac{118}{7000} = .0169 \frac{\text{LBM WATER}}{\text{LBM DRY AIR}}$

SO, WEIGHT OF DRY AIR IS

$1000(1 - .0169) = 983.1$ LBM DRY AIR

a) HEAT REMOVED =

$983.1(41.4 - 20.8) = 20250$ BTU

b) THE SENSIBLE HEAT FACTOR IS .475

$q_{SENSIBLE} = .475(20250) = 9619$ BTU

c) $q_{LATENT} = 20250 - 9619 = 10631$ BTU

d) ABOUT 50.8°F

e) 71.8°F

f) 47F

g) $\omega_2 = 49 \frac{\text{GRAINS}}{\text{LBM}} = .007 \frac{\text{LBM}}{\text{LBM}}$

WEIGHT OF WATER REMOVED,

$983.1 \frac{\text{LBM DRY}}{\text{AIR}}\left[.0169 - .007\right] \frac{\text{LBM WATER}}{\text{LBM DRY AIR}}$

$= 9.71$ LBM WATER

E-I-T HOMEWORK SOLUTIONS: POWER CYCLES

1 THE STEADY FLOW ENERGY EQUATION ON A 1 POUND BASIS IS {EQN 6.104}

$$q - \frac{W}{J} = h_2 - h_1 + \frac{V_2^2 - V_1^2}{2gJ} + z_2 - z_1$$

$q = 0$ {ADIABATIC}
$W = 0$
$z_2 = z_1$
$V_1 = 0$

SO $$\frac{V_2^2}{2Jg} = h_1 - h_2$$

$h_1 = 1279.1$ BTU/LBM FROM P. 6-32
$h_2 = 1243$ FROM P. 6-35

$$V_2 = \sqrt{2Jg(h_1 - h_2)}$$

$$= \sqrt{(2)(778)\frac{\text{FT-LBF}}{\text{BTU}}(32.2)\frac{\text{FT}}{\text{SEC}^2}(1279.1 - 1243)\frac{\text{BTU}}{\text{LBM}}}$$

$$= 1344.8 \text{ FT/SEC}$$

2 FROM PROBLEM 1

$$V_2 = \sqrt{2Jg(h_1 - h_2)}$$

BUT $\Delta h = C_p \Delta T$ {EQN. 6.119}

SO $$V_{IDEAL} = \sqrt{2JgC_p\Delta T}$$

$$= \sqrt{(2)(778)(32.2)(1.9)(1700 - 274)}$$

$$= 11651.2 \text{ FPS}$$

EFFICIENCY IS DEFINED IN TERMS OF ENERGY RATIOS {NOT VELOCITY RATIOS}. IN THIS CASE, KINETIC ENERGY

$$\eta = \frac{E_{K,ACTUAL}}{E_{K,IDEAL}} = \frac{\frac{1}{2} m V_{ACTUAL}^2}{\frac{1}{2} m V_{IDEAL}^2}$$

$$= \frac{(8207)^2}{(11651.2)^2} = .496$$

3 $h_1 = 1472.8$ FROM P. 6-32
$h_2' = 1131.1$ FROM P. 6-31

THE STEAM RATE IS

$$m = \frac{(4000)\text{ HP }(33000)\frac{\text{FT-LBF}}{\text{HP-MIN}}(60)\frac{\text{MIN}}{\text{HR}}}{(778)\frac{\text{FT-LBF}}{\text{BTU}}(1472.8 - 1131.1)\frac{\text{BTU}}{\text{LBM}}} = 29792.1 \text{ LBM/HR}$$

IF THE EXPANSION HAD BEEN ISENTROPIC TO 5 PSIA, THE ENTHALPY WOULD HAVE BEEN

$h_2 = 1077$ {FROM P. 6-35}

$$\eta_{ISEN} = \frac{\Delta h_{ACT}}{\Delta h_{IDEAL}} = \frac{1472.8 - 1131.1}{1472.8 - 1077} = .863$$

4 FROM EQN 7.23

$$\eta = \frac{T_H - T_L}{T_H} = \frac{(650 + 460) - (100 + 460)}{650 + 460} = \frac{550}{1110} = .495$$

5 USE THE PROCEDURE ON PAGE 7-6. REFER TO FIGURE 7.5

AT a: $T_a = 650°F$
$h_a = 696.4$
$s_a = .8833$

AT b: $T_b = 650$
$h_b = 1117.5$
$s_b = 1.2631$

AT c: $T_c = 100°F$
$s_c = s_b = 1.2631$

$$x_c = \frac{1.2631 - .1295}{1.8531} = .612$$

$$h_c = 67.97 + .612(1037.2) = 702.7$$

AT d: $T_d = 100°F$
$s_d = s_a = .8833$

$$x_d = \frac{.8833 - .1295}{1.8531} = .407$$

$$h_d = 67.97 + .407(1037.2) = 490.1$$

DUE TO THE INEFFICIENCIES,

$$h_c' = 1117.5 - .9(1117.5 - 702.7) = 744.2$$
{EQN 7.24a}

$$h_a' = 490.1 + \frac{696.4 - 490.1}{.8} = 748.0$$
{EQN 7.24b}

THEN FROM EQN 7.23

$$\eta_{th} = \frac{(1117.5 - 744.2) - (748.0 - 490.1)}{1117.5 - 748} = .312$$

6 ASSUME THE BOILING WATER IS AT 1 ATM.

$T_{HIGH} = (460 + 212) = 672°R$
$T_{LOW} = 492°R$ {TRIPLE POINT}

$$\eta_{th} = \frac{672 - 492}{672} = .268$$

$$W = \eta Q_{IN} = (.268)(100)\text{BTU} = 26.8 \text{ BTU}$$

$$Q_{OUT} = Q_{IN} - W = 73.2 \text{ BTU}$$

7 FOLLOW THE PROCEDURE ON PAGE 7-7 REFER TO FIGURE 7.7

AT a: $T_a = 360°F$
$P_a = 153.04$ PSIA
$h_a = 332.18$

AT b: $T_b = 360°F$
$h_b = 1194.4$
$S_b = 1.5677$

AT c: $T_c = 100°F$
$S_c = S_b = 1.5677$

$$X_c = \frac{1.5677 - .1295}{1.8531} = .776$$

$$h_c = 67.97 + .776(1037.2) = 872.8$$

AT d: $T_d = 100°F$
$S_d = .1295$
$h_d = 67.97$
$v_f = .01613$
$P_d = .9492$ PSIA

AT e: $P_e = P_a = 153.04$ PSI

$$h_e = h_d + v_f(P_a - P_d)$$

$$= 67.97 + \frac{(.01613)(153.04 - .9492)(144)}{778}$$

$$= 68.42$$

FROM EQN 7.28

$$\eta_{th} = \frac{(1194.4 - 872.8) - (68.42 - 67.97)}{1194.4 - 68.42} = .285$$

8 FIRST ASSUME ISENTROPIC COMPRESSION AND EXPANSION. REFER TO FIGURE 7.7

AT a: $P_a = 100$ PSIA
$h_a = 298.40$

AT b: $P_b = 100$ PSIA
$h_b = 1187.2$
$S_b = 1.6026$

AT c: $P_c = 1$ ATM
$S_c = S_b = 1.6026$

$$X_c = \frac{1.6026 - .3120}{1.4446} = .893$$

$$h_c = 180.07 + .893(970.3) = 1046.5$$

AT d: $T_d = 80°F$
$h_d = 48.02$
$P_d = 14.7$ PSIA
$v_d = .01608$

NOTICE THAT PT. d IS SUB-COOLED AND DOES NOT MATCH FIGURE 7.7 EXACTLY

AT e: $P_e = P_a = 100$ PSIA

$$h_e = 48.02 + \frac{.01608(100 - 14.7)(144)}{778}$$

$$= 48.2$$

NOW, DUE TO INEFFICIENCIES

$$h_c' = 1187.2 - .80(1187.2 - 1046.5) = 1074.6$$

FROM EQN 7.29

$$h_e' = 48.02 + \frac{48.27 - 48.02}{.6} = 48.44$$

FROM EQN 7.30

FROM EQUATION 7.28

$$\eta_{th} = \frac{(1187.2 - 1074.6) - (48.44 - 48.02)}{1187.2 - 48.44}$$

$$= .0985$$

9 USE THE PROCEDURE ON PAGE 7-8 AND REFER TO FIGURE 7.9

AT b: $P_b = 300$ PSIA
$h_b = 393.84$
$S_b = .5879$

AT c: $P_c = 300$ PSIA
$h_c = 1202.8$

AT d: $P_d = 300$
$T_d = 700°F$
$h_d = 1368.3$ (FROM SUPERHEAT TABLES)
$S_d = 1.6751$

AT e: $P_e = 1$ PSIA
$S_e = S_d = 1.6751$

$$X_e = \frac{1.6751 - .1326}{1.8456} = .836$$

$$h_e = 69.7 + .836(1036.3) = 936$$

AT f: $P_f = 1$ PSIA
$h_f = 69.7$
$v_f = .01614$

AT a: $P_a = P_b = 300$

$$h_a = 69.7 + \frac{.01614(300 - 1)144}{778} = 70.6$$

FROM EQN 7.38

$$\eta_{th} = \frac{(1368.3 - 70.6) - (936 - 69.7)}{1368.3 - 70.6} = .332$$

10 REFER TO FIGURE 7.9

AT d: $P_d = 500$
$T_d = 1000°F$
$h_d = 1519.6$
$S_d = 1.7363$

AT e: $P_e = 5$ PSIA
$S_e = S_d = 1.7363$

$$X_e = \frac{1.7363 - .2347}{1.6094} = .933$$

$$h_e = 130.13 + .933(1001) = 1064.1$$

AT f: $h_f = 130.17$

BUT BECAUSE THE TURBINE IS ONLY 75% EFFICIENT

$$h_e' = 1519.6 - .75(1519.6 - 1064.1) = 1178$$

(MORE)

PROBLEM 10, CONTINUED

THE MASS FLOW RATE IS

$$\dot{m} = \frac{(200{,}000)\ \text{KW}\ (1000)\ \frac{\text{W}}{\text{KW}}\ (.05692)\ \frac{\text{BTU}}{\text{MIN-W}}}{(1519.6-1178.0)\ \frac{\text{BTU}}{\text{LBM}}\ (60)\ \frac{\text{SEC}}{\text{MIN}}} = 555.7\ \text{LBM/SEC}$$

$$Q_{out} = (555.7)(1178.0-130.17) = 5.82\ \text{EE5}\ \text{BTU/SEC}$$

11 AT b: $P_b = 200$ PSIA

$T_b = 381.79\ °F$

$h_b = 355.36$

AT c: $h_c = 1198.4$

AT d: $T_d = 400°F$

$P_d = 200$ PSIA

$h_d = 1210.3$

$S_d = 1.5594$

AT e: $P_e = 1.5''\ \text{Hg} = .7365$ PSIA

THIS IS CLOSE TO .6982 PSIA, SO USE 90°F DATA

$S_e = S_d = 1.5594$

$$X_e = \frac{1.5594-.1115}{1.8972} = .763$$

$h_e = 57.99 + .763(1042.9) = 853.7$

AT f: $h_f = 57.99$

$P_f = .7365$

$v_f = .01610$

AT a: $P_a = 200$ PSIA

$$h_a = 57.99 + \frac{(.0161)(200-.7365)144}{778}$$

$= 58.58$

$$\eta_{th} = \frac{(1210.3-58.58)-(853.7-57.99)}{1210.3-58.58} = .309$$

12 USE THE PROCEDURE ON PAGE 7-9 AND REFER TO FIGURE 7.11

AT b: $P_b = 600$ PSIA

$T_b = 486.21\ °F$

$h_b = 471.6$

AT c: $h_c = 1203.2$

AT d: $P_d = 600$ PSIA

$T_d = 700\ °F$

$h_d = 1351.1$

AT e: $P_e = 200$ PSIA

$h_e = 1235$ {FROM MOLLIER DIAGRAM, ASSUMING ISENTROPIC EXPANSION}

$h'_e = 1351.1 - .88(1351.1-1235) = 1248.9$

AT f: $P_f = 200$ PSIA

$T_f = 700°F$

$h_f = 1373.6$

$S_f = 1.7232$

AT g: $T_g = 70°F$

$S_g = S_f = 1.7232$

$$X_g = \frac{1.7232-.0745}{1.9902} = .828$$

$h_g = 38.04 + .828(1054.3) = 911.0$

$h'_g = 1373.6 - .88(1373.6-911) = 966.5$

AT h: $h_h = 38.04$

$P_h = .3631$

$v_h = .01606$

AT a: $P_a = 600$ PSIA

$$h'_a = 38.04 + \frac{(.01606)(600-.3631)144}{.96(778)}$$

$= 39.89$

FROM EQN 8.48

$$\eta_{th} = \frac{(1351.1-39.89)+(1373.6-1248.9)-(966.5-38.04)}{(1351.1-39.89)+(1373.6-1248.9)}$$

$= .353$

13 REFER TO THE FOLLOWING DIAGRAM

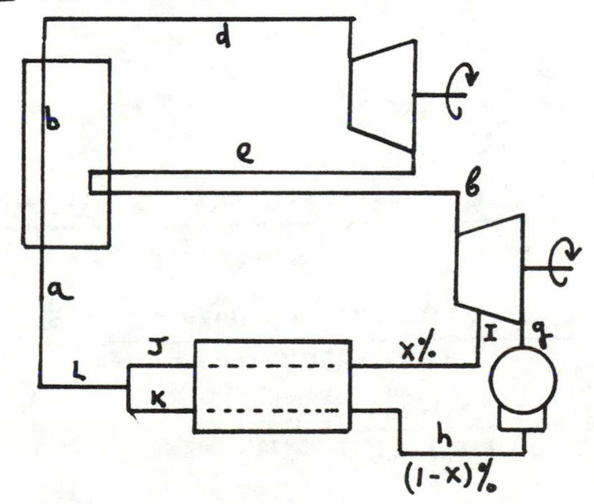

FROM PROBLEM 12

$h_b = 471.6$

$h_d = 1351.1$

$h'_e = 1248.9$

$h_f = 1373.6$

$h_g = 966.5$

$h_h = 38.04$

AT I: THE TEMPERATURE IS 270°F, USING THE MOLLIER DIAGRAM AND ASSUMING ISENTROPIC EXPANSION TO 270°F,

$h_I \approx 1175$

$h'_I = 1373.6 - .88(1373.6-1175) = 1198.8$

AT J: THE WATER IS ASSUMED TO BE SATURATED FLUID AT 270°F.

$h_J = 238.84$

(MORE)

PROBLEM 13 CONTINUED

AT K: THE TEMPERATURE IS $(270-6) = 264°F$ AND SATURATED FLUID.

$h_K = 232.83$

FROM AN ENERGY BALANCE IN THE HEATER,

$(1-x)(h_K - h_h) = x(h_I - h_J)$

$(1-x)(232.83 - 38.04) = x(1198.8 - 238.84)$

$194.79 = x(1154.75)$

$x = .169$

AT L: $h_L = x(h_J) + (1-x)h_K$

$= .169(238.84) + (1-.169)\,232.83$

$= 233.85$

SINCE THIS IS SATURATED LIQUID

$P_L = 38.5$ PSIA

$v_L = .017132$

$T_L = 265°F$

AT a $P_a = 600$ PSIA

$h'_a = 233.85 + \dfrac{.017132(600-38.5)144}{(778)(.96)}$

$= 235.7$

$\eta_{th} = \dfrac{W_{OUT} - W_{IN}}{Q_{IN}}$

$= \dfrac{(h_d - h'_e) + (h_g - h'_I) + (1-x)(h'_I - h_g) - (h'_a - h_L)}{(h_d - h'_a) + (h_g - h'_e)}$

$= \dfrac{(1351.1-1248.9) + (1373.6-1198.8) + (1-.169)(1198.8-966.5)}{1351.1 - 235.7 + 1373.6 - 1248.9}$

$- \dfrac{(235.7-233.85)}{1351.1 - 235.7 + 1373.6 - 1248.9}$

$= \dfrac{468.2}{1240.1} = .378$

14

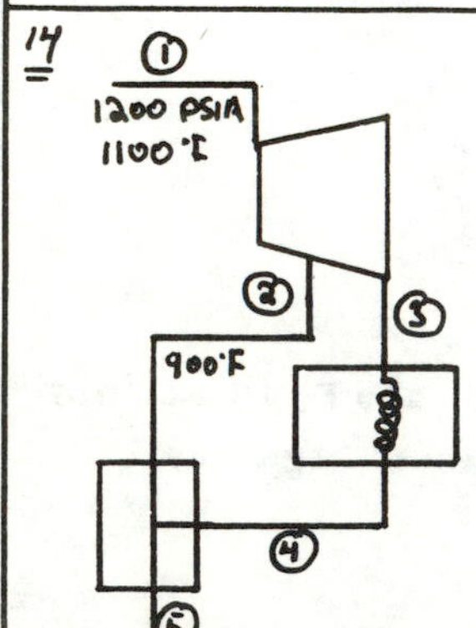

ASSUME ISENTROPIC EXPANSION

ASSUME CONDITION ④ IS SATURATED

NEGLECT PUMP WORK

$h_1 = 1556.9$ $\quad S = 1.6679$

$h_2 = 1460.0$ $\quad S_2 = S_1$

$h_3 = 916.2$ $\quad x = .822$

$h_4 = 59.71$ (SATURATED)

$h_5 = 148$

THE ENERGY BALANCE IN THE HEATER {EQN 7.55} IS

$(1-x)(59.71) + (x)(1460) = 148$

OR $x = .063$

THE WORK OUTPUT {EQN 7.57} IS

$W_{OUT} = .063(1556.9-1460) + (1-.063)(1557-916.2)$

$= 606.5$

THE HEAT INPUT {EQN 7.54} IS

$Q_{IN} = 1556.9 - 148 = 1408.9$

NEGLECTING THE SMALL AMOUNT OF PUMP WORK, THE THERMAL EFFICIENCY IS

$\eta_{th} = \dfrac{606.5}{1408.9} = .43$

15

AT A: $T_a = 560°R$

$h_a = 133.86$ {PAGE 6-36}

$u_a = 95.47$

AT b: $T_b = 820°R$

$h_b = 196.69$

$u_b = 140.47$

AT c: $h_c = h_b + 100 = 296.69$

THIS CORRESPONDS TO 1221°R

$u_c = 212.97$

NOW $\eta_{th} = \dfrac{T_{HI} - T_{LOW}}{T_{HIGH}} = \dfrac{360-100}{360+460} = .317$

$.317 = \dfrac{Q_{IN} - Q_{OUT}}{Q_{IN}} = \dfrac{100 - Q_{OUT}}{100}$

SO $Q_{OUT} = 68.3$

$h_d = h_a + 68.3 = 133.86 + 68.3 = 202.16$

$W_{NET} = Q_{NET} = 100 - 68.3 = 31.7$

$W_{COMP} = h_b - h_a = 196.69 - 133.86 = 62.83$

$W_{EXP} = h_c - h_d = 296.69 - 202.16 = 94.53$

16 REFER TO FIGURE 7.19:

AT a: $V = 11\ FT^3$

$T = 460 + 80 = 540$

$p = 14.3\ PSIA = 2059.2\ PSFA$

$$W = \frac{pV}{RT} = \frac{(2059.2)(11)}{(53.3)(540)} = .787\ LBM$$

AT b: $V_b = \frac{1}{10} V_a = 1.1\ FT^3$

AT c: $$T_c = 540\left(\frac{11}{1.1}\right)^{1.4-1} + \frac{Q_{IN}}{C_V W}$$

$$= 1356.4 + \frac{160}{(.1724)(.787)}$$

$$= 2535.7\ °R = 2075.7\ °F$$

$$\eta_{th} = 1 - \frac{1}{(10)^{1.4-1}} = .602 \quad \{EQN\ 7.78\}$$

17 $$\eta_{th} = 1 - \frac{1}{R^{k-1}} = 1 - \frac{1}{(6)^{1.4-1}} = .512$$

18 ASSUME A 4-STROKE ENGINE. FROM PAGE 7-14, THE NUMBER OF POWER STROKES PER MINUTE IS

$$N = \frac{(2)(3400)(6)}{4} = 10200$$

THEN FROM EQN 7.73,

$$p = \frac{(33000)(79.5)}{\left(\frac{3.125}{12}\right)\frac{\pi}{4}(4)^2(10200)} = 78.6\ PSIG$$

THE TORQUE IS

$$T\ \text{IN FT-LBF} = \frac{(5252)(hp)}{RPM}$$

$$= \frac{(5252)(79.5)}{3400} = 122.8\ FT\text{-}LBF$$

19 THE ACTUAL HORSEPOWER IS

$$hp = \frac{(RPM)(\text{TORQUE IN FT-LBF})}{5252}$$

$$= \frac{(200)(600)}{5252} = 22.86\ HP$$

THE NUMBER OF POWER STROKES PER MINUTE IS

$$N = \frac{(2)(200)(2)}{4} = 200$$

THE STROKE IS $\frac{18}{12} = 1.5\ FT$

THE BORE AREA IS $\frac{\pi}{4}(10)^2 = 78.54\ IN^2$

THE IDEAL HORSEPOWER {FROM EQN 7.73} IS

$$h_p = \frac{(95)(1.5)(78.54)(200)}{33000} = 67.83$$

THE FRICTION HORSEPOWER IS

$$67.83 - 22.86 = 44.97$$

20 USE THE PROCEDURE ON PAGE 7-14:

STEP 1: 1 - 60°F, 14.7 PSIA

2 - 5000 FT ALTITUDE

STEP 2: $IHP_1 = \frac{1000}{.80} = 1250$

STEP 3: $FHP = 1250 - 1000 = 250$

STEP 4: $$\rho_1 = P/RT = \frac{(14.7)(144)}{(53.3)(520)} = .0764\ LBM/FT^3$$

$\rho_2 = .06592$ AT 5000'

STEP 5: $$IHP_2 = 1250\left(\frac{.06592}{.0764}\right) = 1078.5$$

STEP 6: $BHP_2 = 1078.5 - 250 = 828.5$

STEP 7: THE ORIGINAL FLOW RATE OF FUEL IS

$$\dot{W}_{F,1} = BHP_1 BSFC_1 = (1000)(.45)$$

$$= 450\ LBM/HR$$

THE ORIGINAL AIR WEIGHT IS

$$\dot{W}_{A,1} = R_{A/F}(\dot{W}_{F,1}) = (23)(450)$$

$$= 10350\ LBM/HR$$

THIS IS A VOLUME OF

$$V = \frac{wRT}{P} = \frac{(10350)(53.3)(520)}{(14.7)(144)}$$

$$= 1.355\ EE5\ FT^3/HR$$

THIS VOLUME IS THE SAME AT 5000 FT

STEP 8: $\dot{W}_{A,2} = (1.355\ EE5)(.06592) = 8932$

STEP 9: $$\dot{W}_{F,2} = \frac{\dot{W}_{A,2}}{R_{A/F}} = \frac{8932}{23} = 388$$

STEP 10: $$BSFC_2 = \frac{388}{828.5} = .469$$

21 REFER TO FIGURE 7.20:

AT a: $P_a = 14.2$

$T_a = 75°F = 535°R$

FOR 1 POUND OF AIR,

$$V = \frac{(1)(53.3)(535)}{(14.2)(144)} = 13.95\ FT^3$$

AT b: $T_b = 750°F = 1210°R$

$$P_b = (14.2)(144)\left(\frac{1210}{535}\right)^{\frac{1.4}{1.4-1}} = 35576.4\ PSF$$

(MORE)

PROBLEM 21 CONTINUED

$$V_b = \frac{(1)(53.3)(1210)}{35576.4} = 1.813 \text{ FT}^3$$

AT c: $T_c = 2900°F = 3360°R$

$$V_c = (1.813)\frac{3360}{1210} = 5.034$$

AT d: $V_d = V_a = 13.95 \text{ FT}^3$

$$T_d = 3360\left(\frac{5.034}{13.95}\right)^{1.4-1} = 2235.1°R$$

FROM EQN 7.86

$$\eta_{th} = 1 - \frac{2235.1 - 535}{1.4(3360 - 1210)} = .435$$

AT d: $P_d = 14.7 \text{ PSIA} = 2116.8 \text{ PSF}$

$$T_d = 1960\left(\frac{2116.8}{20153}\right)^{\frac{1.4-1}{1.4}} = 1029.5$$

NOW, INCLUDE THE INEFFICIENCIES

$T_a = 520°R$

$$T_b' = 520 + \frac{989.9 - 520}{.83} = 1086°R$$

$T_c = 1960°R$

$$T_d' = 1960 - .92(1960 - 1029.5) = 1103.9$$

FROM EQN 7.94

$$\eta_{th} = \frac{(1960 - 1086) - (1102.9 - 520)}{1960 - 1086} = .332$$

22 REFER TO FIGURE 7.20

AT a: $P_a = 14.7 \text{ PSIA} = 2116.8 \text{ PSF}$

$T_a = 65°F = 525°R$

FOR 1 POUND,

$$V_a = \frac{(1)(53.3)(525)}{2116.8} = 13.22 \text{ FT}^3$$

AT b: $V_b = \frac{V_a}{R} = \frac{13.22}{16} = .8263$

$$T_b = (525)\left(\frac{13.22}{.8263}\right)^{1.4-1} = 1591.5°R$$

AT c: $T_c = 2600°F = 3060°R$

$$V_c = .8263\left(\frac{3060}{1591.5}\right) = 1.589$$

AT d: $V_d = V_a = 13.22$

$$T_d = 3060\left(\frac{1.589}{13.22}\right)^{1.4-1} = 1311$$

FROM EQN 7.86

$$\eta_{th} = 1 - \frac{(1311 - 525)}{1.4(3060 - 1591.5)} = .618$$

23 REFER TO FIGURE 7.23

FIRST, ASSUME ISENTROPIC OPERATION

AT a: $P_a = 14.7 \text{ PSIA} = 2116.8 \text{ PSFA}$

$T_a = 60°F = 520°R$

FOR 1 POUND,

$$V_a = \frac{(1)(53.3)(520)}{2116.8} = 13.09 \text{ FT}^3$$

AT b: $V_b = \frac{13.09}{5} = 2.618$

$$T_b = 520\left(\frac{13.09}{2.618}\right)^{1.4-1} = 989.9°R$$

$$P_b = \frac{(1)(53.3)(989.9)}{2.618} = 20153 \text{ PSF}$$

AT c: $T_c = 1500°F = 1960°R$

$P_c = P_b = 20153 \text{ PSF}$

24 REFER TO FIGURE 7.25

FROM PROBLEM 23,

$T_a = 520°R$

$T_b' = 1086$

$T_d = 1960$

$T_e' = 1103.9$

FROM EQN 7.95 ASSUMING AN IDEAL GAS

$$.65 = \frac{T_c - T_b'}{T_e - T_b'} = \frac{T_c - 1086}{1103.9 - 1086}$$

OR $T_c = 1097.6$

THEN FROM EQN 7.96 ASSUMING AN IDEAL GAS,

$$\eta_{th} = \frac{(1960 - 1103.9) - (1086 - 520)}{1960 - 1097.6} = .336$$

25 a) $COP = \frac{460 + 10}{110 - 10} = 4.7$ {EQN 7.115}

b) $W_{IN} = \frac{1000 \text{ BTU/HR}}{4.7} = 212.7$ {EQN 7.110}

c) $Q_{OUT} = Q_{IN} + W_{IN} = 1000 + 212.7 = 1212.7$

26

$$\frac{460 + 700}{700 - 40} = 1.76$$

27 $Q_{IN} = (100)\text{gpm}\,(.1337)\frac{\text{FT}^3}{\text{GAL}}(62.4)\frac{\text{LBM}}{\text{FT}^3}(1)\frac{\text{BTU}}{\text{LBM-°F}} \times$

$\times\ (80-20)°F = 50057.3 \text{ BTU/MIN}$

$$COP_{IDEAL} = \frac{460 + 20}{80 - 20} = 8$$

$$W_{IN} = \frac{50057.3}{8} = 6257.2 \text{ BTU/MIN}$$

$$\frac{(6257.2)\frac{\text{BTU}}{\text{MIN}}(778)\frac{\text{FT-LBF}}{\text{BTU}}}{(33000)\frac{\text{FT-LBF}}{\text{HP-MIN}}} = 147.5 \text{ HP}$$

28 REFER TO FIGURE 7.32

AT a: SAT. LIQUID

$P_a = 160 \text{ PSIA}$

$h_a = 134.9$ {INTERPOLATED FROM P 6.37}

(MORE)

PROBLEM 28 CONTINUED

AT b: $P_b = 30$ PSIA

$h_b = h_a = 134.9$

AT c: $P_c = 30$ PSIA

$h_c = h_b + 500 = 134.9 + 500 = 634.9$

WHICH IS SUPERHEATED

FROM PAGE 6-38 {WITH P = 30 PSIA AND h = 634.9}

$T \approx 40°F$

$S_c = 1.3845$

AT d: $S_d = S_c = 1.3845$

$P_d = P_a = 160$ PSIA

FROM PAGE 6-38

$T \approx 271.2$

$h \approx 750.1$

$W_{IN} = h_d - h_c = 750.1 - 634.9 = 115.2$

$$(COP)_{HEAT\ PUMP} = \frac{500 + 115.2}{115.2} = 5.34$$

29 REFER TO FIGURE 7.32

AT a: $T_a = 70°F = 530°R$

SAT. LIQUID

$h_a = 23.9$ {FROM P. 6-40}

AT b: $T_b = -30°F = 430°R$

$h_b = h_a = 23.9$

AT c: SAT VAPOR

$T_c = T_b = 430°R$

$h_c = 74.7$

$v_c = 3.088$

$q_{IN} = h_c - h_b = 74.7 - 23.9 = 50.8$ BTU/LBM

$$\dot{m} = \frac{(1)\ \text{TON}\ (200)\ \frac{\text{BTU}}{\text{MIN-TON}}}{(50.8)\ \text{BTU/LBM}} = 3.937\ \text{LBM/MIN}$$

$\dot{V} = (3.937)(3.088) = 12.16$ FT³/MIN

{AT THE COMPRESSOR ENTRANCE}

30 AT a: $T_a = 70°F$

$h_a = 120.5$

AT b: $h_b = h_a = 120.5$

AT c: $T_c = -30°F$

$h_c = 601.4$

$v_c = 18.97$ FT³/LB

$q_{IN} = h_c - h_b = 601.4 - 120.5 = 480.9$ BTU/LBM

$$\dot{m} = \frac{200}{480.9} = .416\ \text{LBM/MIN}$$

$\dot{V} = (.416)(18.97) = 7.89$ FT³/MIN

{AT THE COMPRESSOR ENTRANCE}

31 REFER TO FIGURE 7.33 ASSUME IDEAL GAS

AT c: $T_c = 70°F = 530°R$

$P_c = 14.7$ PSIA

AT d: $P_d = 60$ PSIA

$$T_d = 530\left(\frac{60}{14.7}\right)^{\frac{1.4-1}{1.4}} = 792.1 \text{ IF ISENTROPIC}$$

$$T_d' = 530 + \frac{792.1 - 530}{.7} = 904.5$$

AT a: $T_a = 25°F = 485°R$

$P = 60$ PSIA

AT b: $P_b = P_c = 14.7$ PSIA

$$T_b = 485\left(\frac{14.7}{60}\right)^{\frac{1.4-1}{1.4}} = 324.5°R \text{ IF ISENTROPIC}$$

$$T_b' = 485 - .80(485 - 324.5) = 356.6$$

FROM EQN 7.117a

$$COP = \frac{530 - 356.6}{(904.5 - 485) - (530 - 356.6)} = .708$$

32 FROM EQUATIONS 7.120 AND 7.121

$$R = \frac{65}{14.7} = 4.42$$

$$\eta_v = 1 - \left((4.42)^{\frac{1}{1.33}} - 1\right)(.07) = .856$$

THE WEIGHT OF AIR DISPLACED PER MINUTE IS $\frac{48}{.856} = 56.07$

THE WEIGHT OF THE AIR AND THE CLEARANCE AIR COMPRESSED IS

$(1.07)(56.07) = 60$ LBM/MIN

33 $$\rho = P/RT = \frac{(14.5)(144)}{(53.3)(530)} = .0739\ \text{LBM/FT}^3$$

$$\dot{m} = \frac{(5000)\ \text{FT}^3\text{/HR}\ (.0739)\ \frac{\text{LBM}}{\text{FT}^3}}{(3600)\ \text{SEC/HR}} = .1026\ \text{LBM/SEC}$$

a) FROM P. 6-25

$$W = P_1 V_1 \ln\left(\frac{V_2}{V_1}\right) = wRT_1 \ln\left(\frac{P_1}{P_2}\right)$$

$$P = \frac{W}{t} = \frac{(.1026)(53.3)(530)\ \ln\left(\frac{14.5}{100}\right)}{550}$$

$= 10.18$ HP

b) ASSUME REVERSIBLE ADIABATIC (ISENTROPIC)

$$T_2 = 530\left(\frac{100}{14.5}\right)^{\frac{1.4-1}{1.4}} = 920.2°R$$

$$W = c_v (T_1 - T_2)$$

$$P = \left|\frac{(.1026)\ \frac{\text{LBM}}{\text{SEC}}\ (.1724)\ \frac{\text{BTU}}{\text{LBM-°R}}\ (530 - 920.2)°R\ (778)}{550\ \frac{\text{FT-LBF}}{\text{HP-SEC}}}\right|$$

$= 9.76$ HP

34 ALTHOUGH THE IDEAL GAS LAWS COULD BE USED, IT IS EXPEDIENT TO USE AIR TABLES. FROM PAGE 6-36 AT (460 + 500) = 960

$h_1 = 231.06$

$p_{r,1} = 10.610$

$\phi_1 = .7403$

FOR ISENTROPIC COMPRESSION,

$p_{r,2} = 6 p_{r,1} = 6(10.610) = 63.66$

SEARCHING THE AIR TABLE YIELDS

$T_2 = 1552$

$h_2 = 382.95$

THE ACTUAL ENTHALPY IS

$$h_2' = 231.06 + \frac{382.95 - 231.06}{.65} = 464.74$$

WHICH CORRESPONDS TO 1855°R, AND

$\phi_2' = .91129$

$W = \Delta h = 464.74 - 231.06 = 233.68$

FROM EQUATION 6.29

$$\Delta s = .91129 - .7403 - \left(\frac{53.3}{778}\right) \ln(6)$$

$$= .04824$$

35 INITIALLY: $p_1 = 25$ PSIA

$h_1 = 1160.6$ BTU/LBM

FINALLY: $p_2 = 95$ PSIA

$h_2 = 1279$ {FROM MOLLIER, ASSUMING ISENTROPIC}

$$h_2' = 1160.6 + \frac{1279 - 1160.6}{.70} = 1329.7$$

AT THIS PRESSURE AND ENTHALPY, (INTERPOLATING FROM P. 6-32

$T_2 = 600°F$

$s_2 = 1.7645$

$v_2 = 6.613$

DETAILED TABLES WILL PRODUCE SLIGHTLY DIFFERENT RESULTS

$$W = \frac{(200)\ \frac{\text{LBM}}{\text{MIN}} (1329.7 - 1160.6) \frac{\text{BTU}}{\text{LBM}} (778) \frac{\text{FT-LBF}}{\text{BTU}}}{33000\ \frac{\text{FT-LBF}}{\text{HP-MIN}}}$$

$= 797.3$ HP

E-I-T HOMEWORK SOLUTIONS: CHEMISTRY

2 FROM TABLE 6.3

$R^* = .08206 \frac{\text{ATM-}\ell}{\text{gmole-}^\circ\text{K}}$

3 $pV = WRT = \frac{WR^*T}{(MW)}$

$\frac{(753)\text{MM}(.2875)\ell}{(760)\frac{\text{MM}}{\text{ATM}}} = \frac{(.725)\text{g}(.08206)\frac{\text{ATM-}\ell}{\text{gmole-}^\circ\text{K}}(100+273)}{(MW)}$

$(MW) = (77.9)$ g/gmole

4 FROM $\frac{P_1V_1}{T_1} = \frac{P_2V_2}{T_2}$

$V_2 = \frac{(700)\text{MM}(300)\text{ML}(273+5)^\circ\text{K}}{(273+20)^\circ\text{K}(740)\text{MM}} = 269.3$ ML

5 $\rho = P/RT$

$R = P/\rho T = \frac{(1)\text{ ATM}}{(.178)\text{ g}/\ell\,(273)^\circ\text{K}} = .020579 \frac{\text{ATM-}\ell}{\text{g-}^\circ\text{K}}$

SO $\rho = \frac{(740)\text{ MM/ATM}}{(760)\text{MM/ATM}(.020579)\frac{\text{ATM-}\ell}{\text{g-}^\circ\text{K}}(273+25)^\circ\text{K}}$

$= .15877$ g/ℓ

6 $(MW)_{O_2} = 32$ g/gmole

SO, FROM $pV = WRT$

$T = \frac{(760)\text{MM}(1)\ell(32)\text{ g/gmole}}{(760)\frac{\text{MM}}{\text{ATM}}(1)\text{g}(.08206)\frac{\text{ATM-}\ell}{\text{gmole-}^\circ\text{K}}} = 390^\circ\text{K}$

$= 117^\circ$C

7 $R = P/\rho T = \frac{(1)\text{ ATM}}{(3.22)\frac{\text{g}}{\ell}(273)^\circ\text{K}} = .0011376 \frac{\text{ATM-}\ell}{\text{g-}^\circ\text{K}}$

$W = \frac{PV}{RT} = \frac{(750)\text{MM}(.1)\ell}{(760)\frac{\text{MM}}{\text{ATM}}(.0011376)\frac{\text{ATM-}\ell}{\text{g-}^\circ\text{K}}(273+24)^\circ\text{K}}$

$= .292$ g

8 CALCIUM NITRATE: $Ca(NO_3)_2$

$(MW) = 1(40.1) + 2(14) + 6(16) = 164.1$

SO, $\frac{Ca}{N} = \frac{40.1}{28} = \frac{x}{20}$

$x = 28.64$

9 $(MW)_{CaO} = 1(40.1) + 1(16) = 56.1$

$(MW)_{CaCO_3} = 1(40.1) + 1(12) + 3(16) = 100.1$

$\frac{56.1}{100.1} = .5604$

10 SINCE 1 MOLE OF ANY GAS OCCUPIES 22.4 ℓ AT STP,

$N = \frac{125}{22.4} = 5.58$ moles

11 AT STP, EACH MOLE OCCUPIES 22.4 ℓ, SO

$(MW)_{Cl_2} = 2(35.5) = 71$

$V = \frac{49}{71}(22.4) = 15.46\ \ell$

12 a) $\frac{(57)\text{ g}}{(32)\text{ g/gmole}}(6.023\text{ EE}23) = 1.07\text{ EE}24$

b) $\frac{(15)\ell}{(22.4)\frac{\ell}{\text{gmole}}}(6.023\text{ EE}23) = 4.03\text{ EE}23$

c) $N = \frac{PV}{R^*T} = \frac{(770)\text{MM}(9)\ell}{(760)\frac{\text{MM}}{\text{ATM}}(.08206)\frac{\text{ATM-}\ell}{\text{gmole-}^\circ\text{K}}(273+25)^\circ\text{K}}$

$= .373$ moles

$(.373)(6.023\text{ EE}23) = 2.25\text{ EE}23$

13 AT 22°C (≈ 70°F) THE WATER VAPOR PRESSURE IS .3631 PSI

$h_{\text{MERCURY}} = \frac{P}{\rho} = \frac{(.3631)\text{ LBF/IN}^2}{(.491)\text{ LBM/IN}^3} = .74"$ Hg

$(.74)" = 18.8$ MM

SO $P_{GAS} = 743 - 18.8 = 724.2$ MM Hg

NOW, $PV = \frac{WR^*T}{(MW)}$

$(MW) = \frac{(.1225)\text{g}(.08206)\frac{\text{ATM-}\ell}{\text{gmole-}^\circ\text{K}}(273+22)^\circ\text{K}}{\frac{(724.2)\text{MM}}{(760)\frac{\text{MM}}{\text{ATM}}}(.110)\ell}$

$= 28.3$ g/gmole

14 a) $(MW)_{HBr} = 1.008 + 79.904 = 80.912$

$HBr \longrightarrow H^+ + Br^-$

$(EW) = \frac{80.912}{1} = 80.912$

b) $(MW)_{H_2SO_3} = 2(1.008) + 1(32.06) + 3(16) = 82.076$

$H_2SO_3 \longrightarrow 2H^+ + SO_3^{--}$

$(EW) = \frac{82.076}{2} = 41.038$

c) $(MW)_{H_3PO_4} = 3(1.008) + 1(30.974) + 4(16) = 98.0$

$H_3PO_4 \longrightarrow 3H^+ + PO_4^{---}$

$(EW) = \frac{98.0}{3} = 32.67$

{MORE}

14 CONTINUED

d) $(MW)_{LiOH} = (1)(6.941) + 1(16) + 1(1.008) = 23.95$

$LiOH \longrightarrow Li^+ + OH^-$

$(EW): \frac{23.95}{1} = 23.95$

15

$(MW)_{H_4P_2O_7} = 4(1.008) + 2(30.974) + 7(16) = 177.98$

$H_4P_2O_7 + Na^+ \longrightarrow Na_2H_2P_2O_7 + 2H^+$
177.98 ... 221.94

$\#\text{GRAMS } H_4P_2O_7 = 400\left(\frac{177.98}{221.94}\right) = 320.8$

SINCE 2 HYDROGENS ARE REPLACED, THE EQUIVALENT WEIGHT OF $H_4P_2O_7$ IN THIS REACTION IS

$EW = \frac{177.98}{2} = 88.99$

$\#\text{ EQUIVALENT WEIGHTS} = \frac{320.8}{88.99} = 3.6$

16

a) $\frac{22.99}{1} = 22.99$

b) $\frac{107.868}{1} = 107.87$

c) $\frac{63.546}{2} = 31.773$

17

a) THE SPREAD IN ATOMIC WEIGHTS BETWEEN ELEMENTS 87 AND 118 WILL BE ABOUT 90 {1 PROTON AND 2 NEUTRONS PER ELEMENT CHANGE}

SO $(AW)_{118} \approx 223 + 90 = 313$

b) (MP) INCREASES ABOUT 40°C FOR ELEMENTS 18, 36, 54, 86, SO

$(MP)_{118} \approx -70 + 40 = -30$ (SAY -20 TO -30°C)

c) BOILING POINT IS APPROXIMATELY THE SAME AS MELTING POINT

18 OTHER SALTS OF CHLORINE MADE FROM COLUMN I ARE LiCl, NaCl, AND KCl. THESE ARE WHITE, CRYSTALLINE, IONIC SOLIDS WITH LOW VAPOR PRESSURE, HIGH BOILING POINT, AND WHICH CONDUCT ELECTRICITY WHEN DISSOCIATED.

19 METALLOIDS ARE ELEMENTS WITH PROPERTIES INTERMEDIATE BETWEEN METALS AND NON-METALS, THEY INCLUDE B, Si, Ge, As, Sb, Te, Po

SO - d

20

a) $CH_4 + 2Cl_2 \longrightarrow C + 4HCl$
SINGLE REPLACEMENT (REDOX)

b) $AgNO_3 + HCl \longrightarrow HNO_3 + AgCl$
DOUBLE REPLACEMENT

c) $2AsCl_3 + 3H_2S \rightarrow As_2S_3 + 6HCl$
DOUBLE DISPLACEMENT

d) $2Cu_2O + Cu_2S \longrightarrow 6Cu + SO_2$
REDOX

e) $B_2O_3 + 3Mg \longrightarrow 3MgO + 2B$
REDOX

f) $BaSO_4 + 4C \longrightarrow BaS + 4CO$
REDOX

g) $3Li_2O + P_2O_5 \rightarrow 2Li_3PO_4$
COMBINATION

h) $H_2SO_4 + Ba(OH)_2 \rightarrow 2H_2O + BaSO_4$
DOUBLE DISPLACEMENT

i) $2HNO_3 + CaO \rightarrow Ca(NO_3)_2 + H_2O$
DOUBLE DISPLACEMENT

j) $2H_3PO_4 + 3MgCO_3 \rightarrow Mg_3(PO_4)_2 + 3H_2O + 3CO_2$
DOUBLE DISPLACEMENT WITH DECOMPOSITION

21 USING PAGE 8-18

$2NaCl + H_2SO_4 \rightarrow Na_2SO_4 + 2HCl$

(MW) $117 + 98.1 \rightarrow 142.1 + 73$

THEN

$\frac{Na_2SO_4}{NaCl} = \frac{142.1}{117} = \frac{(.834)X}{(.945)(250)}$

$X = 344$ LBM

22

$\frac{1}{65 \text{ EE-6}} = 1.538 \text{ EE } 4$ LBM

$\text{GAL} = \frac{(1.538 \text{ EE}4)\text{ LBM }(7.48 \frac{\text{GAL}}{\text{FT}^3})}{(62.4)\text{ LBM/FT}^3}$

$= 1844$ GAL

23

$H_2S + 2NH_4OH \longrightarrow (NH_4)_2S + 2H_2O$

(AW) $34.1 + 70 \longrightarrow 68.1 + 36$

a) $\frac{H_2O}{H_2S} = \frac{36}{34.1} = \frac{X}{9}$

$X = 9.501$ g H_2O

$\#\text{ MOLES} = \frac{X}{(MW)} = \frac{9.501}{18} = .528$

b) $\frac{H_2S}{H_2O} = \frac{34.1}{36} = \frac{X}{8}$

$X = 7.578$ g H_2S

$N = \frac{X}{MW} = \frac{7.578}{34.1} = .222$ gmoles

AT STP,

$V = \frac{NRT}{P} = \frac{(.222)(.08206)(273)}{(1)}$

$= 4.97$ ℓ

24 $H_2SO_4 + 2NaOH \rightarrow Na_2SO_4 + 2H_2O$

(AW) $98.1 + 80 \rightarrow 142.1 + 36$

a) 4 MOLES OF WATER $= 4(18) = 72\,g$

$$\frac{H_2SO_4}{H_2O} = \frac{98.1}{36} = \frac{x}{72}$$

$$x = 196.2\ g\ H_2SO_4$$

$$N = \frac{x}{(MW)} = \frac{196.2}{98.1} = 2$$

b) $$\frac{Na_2SO_4}{H_2SO_4} = \frac{142.1}{98.1} = \frac{x}{18}$$

$$x = 26.07\ g\ NaSO_4$$

c) $$\frac{H_2O}{\text{REACTANTS}} = \frac{36}{98.1 + 80} = \frac{x}{100}$$

$$x = 20.2\ g$$

d) 9 g REACTANTS $\longrightarrow$ 9 g PRODUCTS

$$9\left(\frac{142.1}{178.1}\right) = 7.18\ g\ Na_2SO_4$$

$$9\left(\frac{36}{178.1}\right) = 1.82\ g\ H_2O$$

$$N_{Na_2SO_4} = \frac{7.18}{142.1} = .0505$$

$$N_{H_2O} = \frac{1.82}{18} = .1010$$

25 a)

	G	AW	G/AW
Cu	79.9	63.6	1.26
O	20.1	16	1.26

$$\frac{1.26}{1.26} = 1$$

CuO

b)

	G	AW	G/AW
Fe	46.56	55.9	.833
S	53.44	32.1	1.665

$$\frac{.833}{1.665} = .5$$

FeS_2

c)

	G	AW	G/AW
Fe	63.53	55.9	1.136
S	36.47	32.1	1.136

$$\frac{1.136}{1.136} = 1$$

FeS

d)

	G	AW	G/AW
C	85.62	12	7.14
H	14.38	1	14.38

$$\frac{7.14}{14.38} = .496$$

CH_2

e)

	G	AW	G/AW
C	40.0	12	3.33
H	6.7	1	6.7
O	53.3	16	3.33

CH_2O

b) NOTICE THE COMPONENTS ADD UP TO 100% WITHOUT THE H_2O - SO, SOME OF THE HYDROGEN AND OXYGEN MUST BE IN THE FORM OF WATER VAPOR.

$$2H_2 + O_2 \rightarrow 2H_2O$$

(AW) $4 + 32 \rightarrow 36$

OR $\frac{1}{8} + 1 \rightarrow \frac{9}{8}$

SO EACH % OF OXYGEN IN THE FORM OF WATER TAKES 1/8% HYDROGEN. LET X BE THE % OF OXYGEN LOCKED UP IN THE FORM OF WATER, THEN

$$x + \frac{1}{8}x = 7.14$$

$$x = 6.35$$

SO, THE FREE OXYGEN IS

$$57.11 - 6.35 = 50.76$$

THE FREE HYDROGEN IS

$$2.4 - \frac{6.35}{8} = 1.61$$

	G	AW	G/AW
Ca	15.89	40.1	.4
H	1.61	1.0	1.6
P	24.6	31.0	.8
O	50.76	16	3.2
H_2O	7.14	18	.4

$CaH_4P_2O_8 \cdot H_2O$

26 SILVER CHLORIDE (AgCl) IS

$$\frac{107.9}{107.9 + 35.5} = .752 \text{ SILVER BY WEIGHT}$$

THE SILVER CONTENT IS

$$(.752)(7.2) = 5.414$$

SO, THE ORIGINAL PURITY WAS

$$\frac{5.414}{5.82} = .930$$

27 $CuSO_4 \cdot 5H_2O$ IS

$$\frac{63.6}{63.6 + 32.1 + 4(16) + 5(18)} = .2547 \text{ COPPER}$$

THE COPPER CONTENT IS

$$(.2547)(30) = 7.64$$

THE PURITY IS

$$\frac{7.64}{10} = .764$$

28 # g moles of $CH_4 = \frac{200}{22.4} = 8.929$

1 mole of methane has a molecular weight of 16 g, so, the methane has a mass of

$(16)(8.929) = 142.86$ g

In pounds,

$(142.86)\text{g}(.0022046)\text{LB/g} = .315$ LB

The available heat is

$(.315)(24{,}000)$ BTU/LB $= 7560$ BTU

$\Delta T = (95-15)\left(\frac{9}{5}\right) = 144°F$

$q = Mc_p\Delta T/m$

$$M = \frac{(.5)(7560)(.4536)\ \text{kg/lbm}}{(1)(144)} = 11.91\ \text{kg}$$

29 $C_3H_8 + 5O_2 \rightarrow 3CO_2 + 4H_2O$

(MW) $44 + 160 \rightarrow 132 + 72$

$$\frac{CO_2}{C_3H_8} = \frac{132}{44} = \frac{X}{15}$$

$X = 45$ LBM/HR

$$V = \frac{WRT}{P} = \frac{(45)(35.1)(530)°R}{(14.7)(144)} = 395.5\ \text{FT}^3/\text{HR}$$

30 $CH_4 + 2O_2 \rightarrow CO_2 + 2H_2O$

(volumes) 1 2 1 2

The ideal oxygen volume is $(2)(4000) = 8000$ CFH.
The actual oxygen volume is $(1.3)(8000) = 10{,}400$ CFH
From page 6-14, assuming air is 20.9% oxygen by volume, the oxygen weight is

$$m = \frac{pV}{RT} = \frac{(15)(144)(0.209)(10{,}400)}{(48.3)(460+100)} = 173.6$$

But air is .2315 oxygen by weight, so nitrogen weight is

$$\frac{1-.2315}{.2315}(173.6) = 576.3\ \text{LBM/HR}$$

31 $C + O_2 \rightarrow CO_2$

$12 + 32 \rightarrow 48$

so $\frac{32}{12} = 2.67$ LBM oxygen req'd per pound carbon

$2H_2 + O_2 \rightarrow 2H_2O$

$4 + 32 \rightarrow 36$

so $\frac{32}{4} = 8\ \frac{\text{LBM } O_2}{\text{LBM } H_2}$

$S + O_2 \rightarrow SO_2$

$32.1 + 32 \rightarrow 64.1$

so $\frac{32}{32.1} = 1\ \frac{\text{LBM } O_2}{\text{LBM } S}$

Nitrogen does not burn.

$$(.84)(2.67) + (.153)(8) + (.003)(1) = 3.47\ \frac{\text{LBM } O_2}{\text{LBM FUEL}}$$

But air is .2315 oxygen. So,

$$\text{AIR} = \frac{3.47}{.2315} = 15\ \frac{\text{LBM AIR}}{\text{LBM FUEL}}$$

32 Ideally,

$C_3H_8 + 5O_2 \rightarrow 3CO_2 + 4H_2O$

$44 + 160 \rightarrow 132 + 72$

The excess oxygen is $(160)(.2) = 32$
Since air is .2315 oxygen by weight, the nitrogen is

$$\frac{(1-.2315)}{.2315}(160+32) = 637.4$$

% CO_2 by weight =

$$\frac{132}{132+72+32+637.4} = .151\ \text{(WET)}$$

33 Write the oxidation numbers for each element:

a) $H_2 + Cl_2 \rightarrow 2HCl$

H = 0, Cl = 0 → H = +1, Cl = −1

H has become less negative → oxidized
Cl has become more negative → reduced

b) $Zn + H_2SO_4 \rightarrow H_2 + ZnSO_4$

Zn = 0, H = +1, SO_4 = −2 → H = 0, Zn = +2, SO_4 = −2

Zn oxidized
H reduced

c) $2KBr + Cl_2 \rightarrow Br_2 + 2KCl$

K = +1, Br = −1, Cl = 0 → Br = 0, K = +1, Cl = −1

Br oxidized
Cl reduced

34 a) $3Zn + N_2 \rightarrow Zn_3N_2$

b) $2FeCl_3 + SnCl_2 \rightarrow 2FeCl_2 + SnCl_4$

35 a) $2FeCl_2 + Cl_2 \rightarrow 2FeCl_3$

b) $H_2SO_4 + 2KMnO_4 \rightarrow K_2SO_4 + Mn_2O_7 + H_2O$

c) $2Al + 3H_2SO_4 \rightarrow Al_2(SO_4)_3 + 3H_2$

d) $Fe + Ag_2SO_4 \rightarrow FeSO_4 + 2Ag$
(Since iron is a stronger reducer than silver.)

36 N_2, CO_2 don't burn

$2CO + O_2 \rightarrow 2CO_2$

$2H_2 + O_2 \rightarrow 2H_2O$

(more)

PROBLEM 36 CONTINUED

SO $(.06)CO_2 + (.22)CO + (.12)H_2 + (.60)N_2 + (.17)O_2$
$\longrightarrow (.60)N_2 + (.28)CO_2 + .12(H_2O)$

ASSUME THE WATER IS A VAPOR. USE THE ENTHALPIES OF FORMATION IN TABLE 8.6

$(.06)(-94.05) + (.22)(-26.42) + (.12)(0) + (.60)(0) + (.17)(0) \longrightarrow (.60)(0) + (.28)(-94.05) + (.12)(-57.80)$

THEN $\sum \text{PRODUCTS} - \sum \text{REACTANTS} = -21.81 \ \frac{\text{kcal}}{\text{mole Fuel}}$

SINCE 1 MOLE AT STP OCCUPIES 22.4 ℓ,

HEATING VALUE $= \frac{21.81}{22.4} = .974 \ \frac{\text{kcal}}{\ell}$

37 $2CO + 1.25O_2 \longrightarrow 2CO_2 + .25O_2$
FROM TABLE 8.6

$2(-26.42) + 1.25(0) \longrightarrow 2(-94.05) + .25(0)$

$\frac{\sum \text{PRODUCTS} - \sum \text{REACTANTS}}{2} = -67.63 \ \frac{\text{kcal}}{\text{mole CO}}$

THE MOLECULAR WEIGHT OF CO_2 IS $12+32=44$. THE WEIGHT OF CO_2 PRODUCED IS $2(44)=88$. SIMILARLY, THE WEIGHT OF O_2 LEFT OVER IS 8 g. BECAUSE AIR IS 79.1% NITROGEN BY VOLUME, THERE ARE ALSO

$\left(\frac{79.1}{100-79.1}\right) 1.25 = 4.73$ moles OF NITROGEN.

NITROGEN 'WEIGHS' 28 g/gmole, SO THE WEIGHT OF NITROGEN IS

$(4.73)(28) = 132.44$ g

AS AN APPROXIMATION, USE TABLE 6.4 DATA AND ASSUME ALL COMBUSTION HEAT GOES INTO THE FLUE GASES.

$T_{FINAL} = T_{INITIAL} + \frac{\text{HEAT OF COMBUSTION}}{\sum C_p M}$

ASSUME $T_{INITIAL} = 20°C$

$T_{FINAL} = 20 + \frac{(67.63)\frac{\text{kcal}}{\text{mole}}(1000)\frac{\text{cal}}{\text{kcal}}(2)\text{moles}}{(88)(.205) + (8)(.217) + (132.44)(.247) \ \frac{\text{cal}}{\text{g-°C}}}$

$= 2597°C$

38 SUCROSE IS $C_{12}H_{22}O_{11}$
THE MOLECULAR WEIGHT IS
$12(12) + 22(1) + 11(16) = 342$
THE SOLVENT WEIGHT IS
SOLUTION WEIGHT − SOLUTE WEIGHT
$= (1.05)(100) - 13.5 = 91.5$

$M = \frac{(13.5)/(342)}{.100} = .395$

$\frac{13.5}{91.5} = \frac{x}{1000} \qquad x = 147.54$

$m = \frac{147.54}{342} = .431$

39 ACETIC ACID IS CH_3COOH
$(MW) = 2(12) + 4(1) + 2(16) = 60$
$\#\text{gmoles} = \frac{10}{60} = .1667$

a) $m = \frac{.1667}{.125} = 1.334$

b) # OF gmoles OF WATER $= \frac{125}{18} = 6.944$

MOLE FRACTION ACID $= \frac{.1667}{.1667 + 6.944} = .023$

MOLE FRACTION WATER $= \frac{6.944}{.1667 + 6.944} = .977$

40 $(MW)_{NaOH} = 23 + 16 + 1 = 40$,
$(gew) = \frac{40}{1} = 40$

THE NUMBER OF GRAMS NEEDED IS
$(60)\ \ell\ (40)\frac{g}{gew}(.5)\frac{gew}{\ell} = 1200 g$

NEEDED: $\frac{(1200)\ g}{(1000)\ g/kg\ (1-.12)} = 1.364$

41 THE WEIGHT OF 55 GALLONS IS

$\frac{(1.0835)(62.43)\frac{\text{LBM}}{\text{FT}^3}(55)\text{ GAL}}{(7.48)\frac{\text{GAL}}{\text{FT}^3}} = 497.4$ LBM

THIS IS 10% BY WEIGHT SOLUTE, SO
$(.10)(497.4) = 49.74$ LBM

42 a) $(MW) = 2(1) + 1(32.1) + 4(16) = 98.1$ g/gmole
$\#\text{GRAMS} = (2.5)(2)(98.1) = 490.5$ g

b) $(MW) = 1(137.3) + 2(16) + 2(1) = 171.3$ g/gmole
$\#\text{GRAMS} = (5)(.525)(171.3) = 449.7$ g

c) $(MW) = 2(27.0) + 3(32.1) + 12(16) = 342.3$
$(gew) = \frac{342.3}{6} = 57.05$
$\#\text{GRAMS} = (.350)(.5)(57.05) = 9.98$

d) $(MW) = 1(39.1) + 1(54.9) + 4(16) = 158$
$(gew) = \frac{158}{1} = 158$
$\#\text{GRAMS} = (.025)(1.9)(158) = 7.51$

43 $pH = -\log_{10}[H^+]$
$= -\log[3\ EE{-3}] = 2.52$

44 $[H_3O^+] = \text{ANTILOG}_{10}(-6.5) = 3.162\ EE{-7}$

$pOH = 14 - pH = 14 - 6.5 = 7.5$

$[OH^-] = \text{ANTILOG}_{10}(-7.5) = 3.16\ EE{-8}$

45 .5N MEANS THERE ARE $\frac{1}{2}$(gew) PER LITER. IN THIS CASE (MW) = (gew), SO THERE ARE $\frac{1}{2}$ MOLES PER LITER

$$[H_3O^+] = (.93)(.5) = .465$$

$$pH = -LOG(.465) = .333$$

46 $pH = -LOG[5\ EE\text{-}6] = 5.3$

$pOH = 14 - 5.3 = 8.7$

47 THE REACTION IS

$$HC_2H_3O_2 + NaC_2H_3O_2 + H_2O \rightarrow H_3O^+ + C_2H_3O^- + Na^+$$

THE IONIZATION CONSTANT FOR THE ACETATE ION IS UNCHANGED. THAT IS

$$\frac{[H_3O^+][C_2H_3O_2^-]}{[HC_2H_3O_2]} = 1.8\ EE\text{-}5$$

NOW, $[H_3O^+]$ IS UNKNOWN. LET IT BE X

$[C_2H_3O^-]$ COMES FROM THE ACETIC ACID {1 FOR 1 WITH $[H_3O^+]$} AND FROM $NaC_2H_3O_2$

SO

$$[C_2H_3O_2^-] = X + (M)_{NaC_2H_3O_2}(\%\ IONIZED)$$

$$= X + (.01)(1) = X + .01$$

SINCE H_3O^+ RESULTS ONLY FROM IONIZATION OF ACETIC ACID {1 FOR 1}

$$[HC_2H_3O_2] = (M)_{HC_2H_3O_2} - [H_3O^+]$$

$$= .01 - X$$

SO $\frac{(X)(X+.01)}{.01-X} = 1.8\ EE\text{-}5$

BUT X IS VERY SMALL.

$X + .01 \approx .01$

$.01 - X \approx .01$

SO $X = 1.8\ EE\text{-}5$

$pH = -LOG(X) = 4.74$

48 $[H_3O^+] = ANTILOG(-5) = .00001$

NOW [ ] = (M)(FRACTION IONIZED)

SO, FRACTION IONIZED $= \frac{.00001}{.005} = .002$

OR .2%

49 $NH_3 + H_2O \longrightarrow NH_4^+ + OH^-$

$$K_{ION} = \frac{[NH_4^+][OH^-]}{[NH_3]}$$

$[NH_4^+] = (1)\frac{mole}{l}(.004) = .004$

$[OH^-] = [NH_4^+] = .004$

$[NH_3] = (M)(\%\ IONIZED)$

$= (1)(.996) = .996$

$$K_{ION} = \frac{(.004)(.004)}{.996} = 1.606\ EE\text{-}5$$

50 FROM PROBLEM #49, $K_{ION} = 1.606\ EE\text{-}5$

SO $\frac{[NH_4O^+][OH^-]}{[NH_3]} = 1.606\ EE\text{-}5$

$[NH_4O^+] = [OH^-] = X$ {UNKNOWN}

$[NH_3] = .05 - X \approx .05$

SO $\frac{X^2}{.05} = 1.606\ EE\text{-}5$

$X = 8.96\ EE\text{-}4$

51 $\Delta T_f = 12$

$K_{f\,(WATER)} = 1.86$

FROM EQN 8.20

$$m = \frac{\#g\,moles}{\left(\frac{g\ solvent}{1000}\right)} = \frac{\Delta T_f}{K_f}$$

$$\#g\,moles = \frac{(12)(20)(1000)\ g/l}{(1.86)(1000)} = 129.03$$

METHYL ALCOHOL IS CH_3OH

$(MW) = 1(12) + 4(1) + 16 = 32$

$WT = (32)(129.03) = 4129\ g$

OR 4.13 kg

52 GLYCOL {ETHYLENE GLYCOL} IS $(CH_2OH)_2$

$(MW) = 2(12) + 6(1) + 2(16) = 62$

$$m = \frac{\left(\frac{(5)(1000)}{62}\right)}{15} = 5.376$$

FROM EQN 8.20

$\Delta T_f = (1.86)(5.376) = 10°C$

53 $K_f = 5.12$

$T_f = 5.5°C$

$\Delta T_f = 5.5 - 4.95 = .55$

$$m = \frac{\left(\frac{1.15}{(MW)}\right)}{.1} = \frac{11.5}{MW}$$

FROM EQN 8.20

$.55 = \left(\frac{11.5}{MW}\right)(5.12)$

$MW \approx 107.05$

54 2 qts = $\frac{1}{2}$ GALLON

$$(.5)\,gal\,(.1337)\frac{FT^3}{GAL}(28.32)\frac{l}{FT^3}(1000)\,g/l = 1893.2\ g$$

{MORE}

PROBLEM # 54 CONTINUED

$(MW)_{NaCl} = 23 + 35.5 = 58.5$

$$m = \frac{\left(\frac{5}{58.5}\right)}{1.893} = .04515$$

FROM TABLE 8.4 $K_b = .512$. BUT COMPLETE IONIZATION INTO TWO IONS DOUBLES THE MOLECULAR IONIZATION. SO, FROM EQUATION 8.20,

$$\Delta T_b = (.512)(2)(.04515) = .046°C$$

55 FROM EQN 8.25

$(65.8)(3) = (50)(N)_{Na_2CO_3}$

$(N)_{Na_2CO_3} = 3.948$

$(MW) = 2(23) + 12 + 3(16) = 106$

$gew = \frac{106}{2} = 53$

CONSIDER 1 ℓ OF Na_2CO_3 SOLUTION:

WT OF SOLUTE $= (3.948)(53) = 209.2$.

THE TOTAL SOLUTION WEIGHT IS

$(1.25)(1000) = 1250$

$$G_{Na_2CO_3} = \frac{209.2}{1250} = .167$$

56 FROM EQUATIONS 8.25 AND 8.26,

NORMALITY = (FORMALITY)(Δ_{CHARGE})

ASSUME COMPLETE IONIZATION

$(20)(F_{H_2SO_4})(2) = (38.3)(.103)(1)$

$F = .0986$

57 IN THIS CASE, $f = 1$, SO $N = F$

$(25)(F_{HCl}) = (24.3)(.1035)$

$F_{HCl} = .1006$

58 AW OF Ag IS 107.9

$gEW = 107.9/1 = 107.9$

$\# FARADAYS = \frac{2}{107.9}$

AW OF Sn IS 118.7

$gEW = 118.7/2 = 59.35$

FROM EQN 8.41

$$\# grams = \left(\frac{2}{107.9}\right)(59.35) = 1.1$$

59 (AW) OF ALUMINUM = 27

$$(gew) = \frac{2(27)}{6} = 9$$

FROM EQN 8.41

$$(100)g = \frac{(125)\,Amps\,(t)\,sec}{(96{,}500)\,amp\text{-}sec}\,(9)\,g$$

$t = 8578$ sec

60 $$\# FARADAYS = \frac{(2.5)\,AMP\,(3600)\,sec}{(96{,}500)\,AMP\text{-}sec} = .09326$$

$$(gew)_{H_2} = \frac{2}{2} = 1$$

GRAMS OF $H_2 = (1)(.09326) = .09326$

FROM $PV = \frac{wR^*T}{(MW)}$

$$V = \frac{(.09326)(.08206)(273+25)}{(2)\left(\frac{780}{760}\right)} = 1.111\ \ell$$

$$(gew)_{O_2} = \frac{32}{4} = 8$$

GRAMS OF $O_2 = (8)(.09326) = .7461$

$$V = \frac{(.7461)(.08206)(298)}{(32)\left(\frac{780}{760}\right)} = .556\ \ell$$

TOTAL VOLUME {IGNORING WATER VAPOR}

$1.111 + .556 = 1.667\ \ell$

61 $(MW)_{C_2H_4Br_2} = 2(12) + 4(1) + 2(79.9) = 187.8$

$$(\# moles)_{C_2H_4Br} = \frac{10}{187.8} = .0532$$

$(MW)_{C_3H_6Br_2} = 3(12) + 6(1) + 2(79.9) = 201.8$

$$(\# moles)_{C_3H_6Br_2} = \frac{80}{201.8} = .3964$$

SINCE THE PARTIAL PRESSURE IS PROPORTIONAL TO THE MOLE FRACTION,

$$(PP)_{C_2H_4Br} = (173)\frac{.0532}{.0532 + .3964} = 20.47\ mm\ Hg$$

$$(PP)_{C_3H_6Br} = (127)\frac{.3964}{(.0532 + .3964)} = 111.98\ mm\ Hg$$

62 $(MW)_{H_2O} = 2(1) + 16 = 18$

$$(\# moles)_{H_2O} = \frac{500}{18} = 27.78$$

LET X = MW OF SUCROSE

$$(\# moles)_{SUCROSE} = \frac{57}{X}$$

BY RAOULTS LAW,

$$(.092) = (17.54)\frac{57/X}{27.78 + \frac{57}{X}}$$

$X = 389.2$

63 THE SOLUBILITY MAY ALSO BE WRITTEN AS 14 EE-3 g/ℓ

$(MW) = 1(137.3) + 1(12) + 3(16) = 197.3$

$$SO,\ \frac{\# moles}{\ell} = \frac{14\ EE\text{-}3}{197.3} = 7.096\ EE\text{-}5$$

{MORE}

PROBLEM #63 CONTINUED

$K_{sp} = [Ba][CO_3]$

$= (7.096\ EE\text{-}5)^2 = 5.035\ EE\text{-}9$

64 $PbF_2 \rightarrow Pb^{++} + 2F^-$

$K_{sp} = [Pb^{++}][F^-]^2$

BUT $[F^{--}] = 2[Pb^{++}]$

LET $[Pb^{++}] = x$

$3.2\ EE\text{-}8 = x(2x)^2$

$x = 2\ EE\text{-}3\ \frac{g\text{-ions}}{l}$

$[F^-] = 4\ EE\text{-}3$

65 $Ag_2CrO_4 \longrightarrow 2Ag^+ + CrO_4^{--}$

$[Ag^+]^2[CrO_4^{--}] = 1.1\ EE\text{-}12$

THE $[Ag^+]$ CONTRIBUTED BY THE Ag_2CrO_4 IS INSIGNIFICANT COMPARED TO THAT FOR THE $AgNO_3$

$[Ag^+] = (.5)\ \frac{g\text{-ions}}{l}$

THEN $[CrO_4^{--}] = \frac{1.1\ EE\text{-}12}{(.5)^2} = 4.4\ EE\text{-}12$

E-I-T HOMEWORK SOLUTIONS: STATICS

1 THE BLOCK'S FREEBODY IS

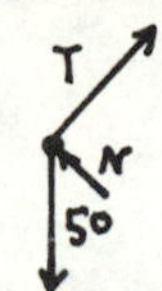

13, 12, 5

THE ANGLE TRIANGLE IS

RESOLVING THE 50 POUNDS INTO COMPONENTS PARALLEL AND PERPENDICULAR TO THE PLANE GIVES

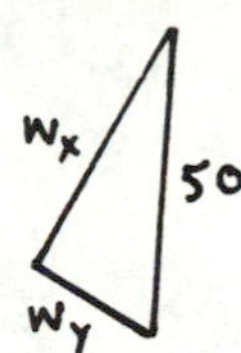

BY SIMILAR TRIANGLES, $W_x = \frac{12}{13}(50) = 46.15$

SO $T = 46.15$

2 FOR THE INCLINED LOAD

$F_x = \frac{3}{5}(100) = 60$

$F_y = \frac{4}{5}(100) = 80$

a) SO M_A (+): $600(4) - 80(9) = 1680$ FT-LBF, CW

b) $\Sigma F_x = 200 + 60 = 260$

$\Sigma F_y = (80 - 600) = -520$

$R = \sqrt{(260)^2 + (-520)^2} = 581.4$ LBF

$\theta = \arctan\left(\frac{-520}{260}\right) = -63.43$

LOCATION $= \frac{1680}{520} = 3.23$ FT FROM PT A

3 $\Sigma F_x \overset{+}{\rightarrow}$: $100\left(\frac{\sqrt{3}}{2}\right) - 200(\cos 30°) = -102.5$

$\Sigma F_y + \uparrow$: $300 + 100\left(\frac{\sqrt{3}}{2}\right) + 200(\sin 30°) = 470.7$

ΣM (+): $300(2) - 100\left(\frac{\sqrt{3}}{2}\right)2 - 200(2)$

$= 58.6$ FT-LBF CCW

4

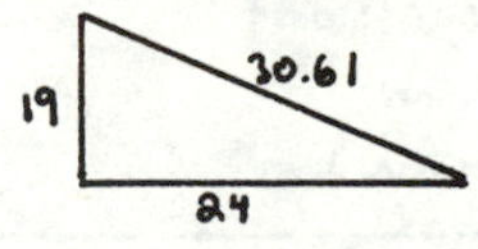

$1000\left(\frac{19}{24}\right) = 791.6$

$\left(\frac{30.61}{24}\right)1000 = 1275.4$

1000

5 ΣM_A (+): $B_y(4) - 200(6) + 500 = 0$

$B_y = +175$ {UPWARDS}

ΣM_B (+): $-A_y(4) + 200(6) - 500 = 0$

$A_y = 175$ (DOWNWARD) $A_x = -200$ TO LEFT

6 TOTAL DISTRIBUTED LOAD

$\left(\frac{1}{2}\right)(18)(100) = 900$ LBF

THE RESULTANT ACTS $\frac{2}{3}(18) = 12$ UP FROM POINT A. SO

ΣM_A (+): $900(12) - T(18) = 0$

$T = -600$ IN CORD

ΣM_{tip} (+): $-900(6) + A(18) = 0$

$A = -300$

7 $B_x = \frac{4}{5}B$

$B_y = \frac{3}{5}B$

ΣM_C (+): $(-100)(8) - B_y(10) - B_x(8) + 500 = 0$

$= -800 - (3/5)B(10) - (4/5)B(8) + 500 = 0$

OR $B = 24.19$

$B_x = \left(-\frac{4}{5}\right)(24.19) = -19.35$ {TO THE LEFT}

$B_y = \left(\frac{3}{5}\right)(24.19) = 14.51$ {DOWN}

8a ΣM_C (+): $D_y(6) - 8000(6) + 1600(16) = 0$

$D_y = 3733$

SINCE DE IS THE ONLY VERTICAL MEMBER LEAVING POINT D,

DE = 3733 {COMPRESSION}

8b BY SYMMETRY, $A_y = L_y = 160$ KIPS

FOR DE

CUT AS SHOWN AND SUM VERTICAL FORCES.

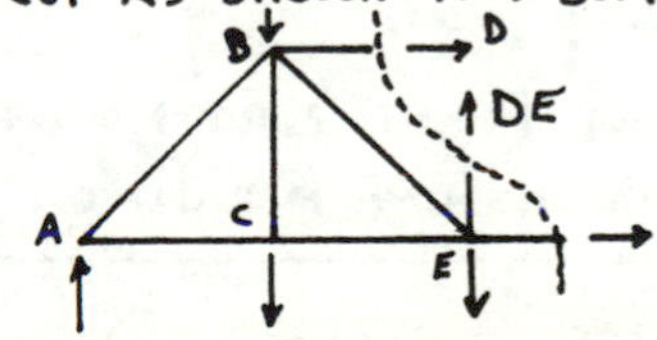

$\Sigma F_y + \uparrow$: $160 + DE - 60 - 60 - 4 = 0$

DE = -36 KIPS {COMPRESSION}

FOR HJ

CUT AS SHOWN AND TAKE MOMENTS ABOUT I

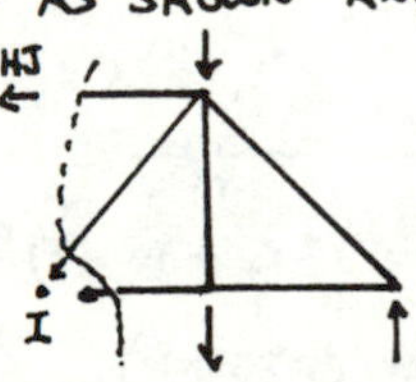

ΣM_I (+): $160(60) - 60(30) - 4(30) + HJ(20) = 0$

HJ = -384 {COMPRESSION}

9 AT PIN F

FG = 0

FN = 2000

AT PIN E

$\Sigma F_y = 0 + \uparrow$:

$ED_y - 5000 = 0$

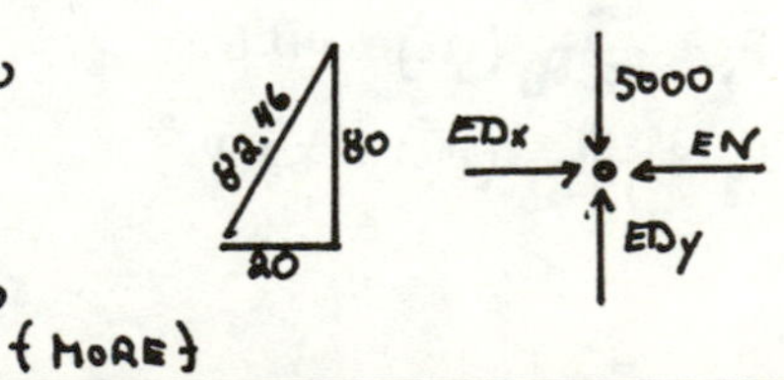

{MORE}

PROBLEM # 9 CONTINUED

so $ED_y = 5000$

$$ED_x = \frac{20}{80}(5000) = 1250$$

$\Sigma F_x = 0 \xrightarrow{+}: \quad 1250 - EN = 0$

$EN = 1250$

AT PIN N

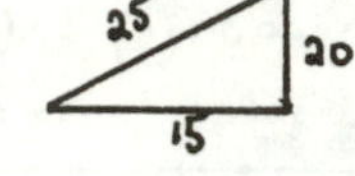

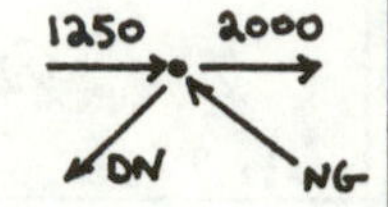

$\Sigma F_y + \uparrow = 0:$

$NG_y - DN_y = 0$

OR $NG_y = DN_y$

AND SINCE THE ANGLES ARE THE SAME,

$NG = DN$

$\Sigma F_x \xrightarrow{+} = 0: \quad 2000 + 1250 - 2(.6)DN = 0$

$DN = NG = 2708.3$

AT PIN G

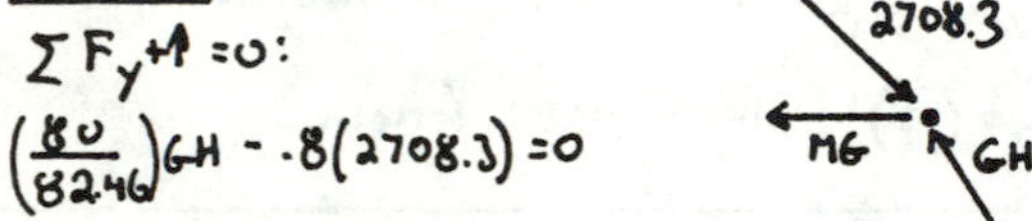

$\Sigma F_y + \uparrow = 0:$

$$\left(\frac{80}{82.46}\right)GH - .8(2708.3) = 0$$

$GH = 2233.3$

$$GH_x = \left(\frac{20}{82.46}\right)GH = 541.67$$

$\Sigma F_x = 0 \xrightarrow{+}: \quad .6(2708.3) - 541.67 - MG = 0$

$MG = 1083.3$ T

10 CUT AND SUM MOMENTS ABOUT A

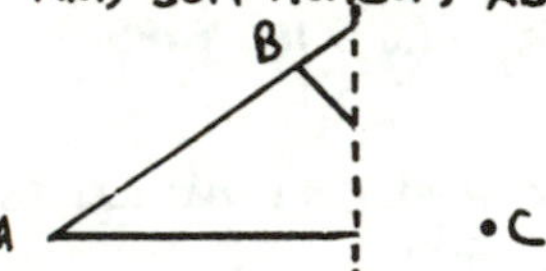

ALL FORCES DROP OUT (HAVE ZERO MOMENT ARM) EXCEPT BC. WITH NOTHING TO BALANCE IT, BC = 0

11 BY SYMMETRY,

$$A_y = \tfrac{1}{2}(300) = 150$$

CUT AND SUM MOMENTS ABOUT D

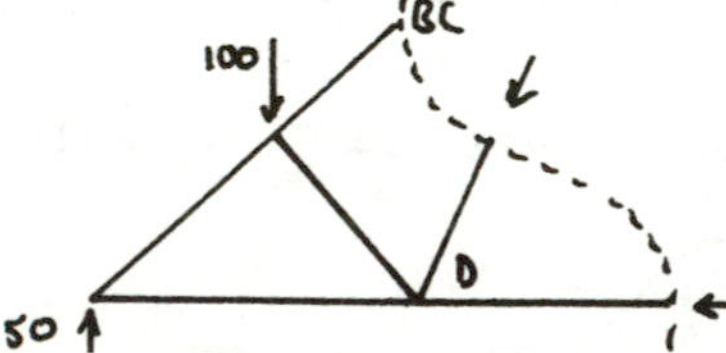

$$\Sigma M_D \circlearrowleft: (5)\left(\frac{\sqrt{2}}{2}\right)BC + (5)\left(\frac{\sqrt{2}}{2}\right)BC - 100(5) + 150(10) = 0$$

$BC = -141.4$ (COMPRESSION)

12

$$\sqrt{(7)^2 + (9)^2 + (4)^2} = 12.08$$

$$F_x = \frac{7}{12.08}(190) = 110.1$$

$$F_y = \frac{9}{12.08}(190) = 141.6$$

$$F_z = \frac{4}{12.08}(190) = 62.9$$

13 USE THE PROCEDURE ON PAGE 9-11. BUT FIRST, MOVE THE ORIGIN TO THE APEX OF THE TRIPOD SO THAT THE COORDINATES OF THE TRIPOD BASES BECOME

A = (5, -12, 0)
B = (0, -8, -8)
C = (-4, -7, 6)

BY INSPECTION,

$F_x = 1200 \quad F_y = 0 \quad F_z = 0$

$$L_A = \sqrt{(5)^2 + (-12)^2 + (0)^2} = 13$$

$$\cos\theta_{xA} = \frac{5}{13} = .385$$

$$\cos\theta_{yA} = \frac{-12}{13} = -.923$$

$$\cos\theta_{zA} = \frac{0}{13} = 0$$

$$L_B = \sqrt{(0)^2 + (-8)^2 + (-8)^2} = 11.31$$

$$\cos\theta_{xB} = \frac{0}{11.31} = 0$$

$$\cos\theta_{yB} = \frac{-8}{11.31} = -.707$$

$$\cos\theta_{zB} = \frac{-8}{11.31} = -.707$$

$$L_C = \sqrt{(-4)^2 + (-7)^2 + (6)^2} = 10.05$$

$$\cos\theta_{xC} = \frac{-4}{10.05} = -.398$$

$$\cos\theta_{yC} = \frac{-7}{10.05} = -.697$$

$$\cos\theta_{zC} = \frac{6}{10.05} = .597$$

FROM EQUATIONS 9.65, 9.66, + 9.67

$$.385\,F_A \qquad\qquad -.398\,F_C = -1200$$

$$-.923\,F_A - .707\,F_B - .697\,F_C = 0$$

$$-.707\,F_B + .597\,F_C = 0$$

SOLVING THESE SIMULTANEOUSLY YIELDS

$F_A = -1793$ (COMPRESSION)
$F_B = 1080$ (TENSION)
$F_C = 1279$ (TENSION)

14 IN THE X-Y PLANE

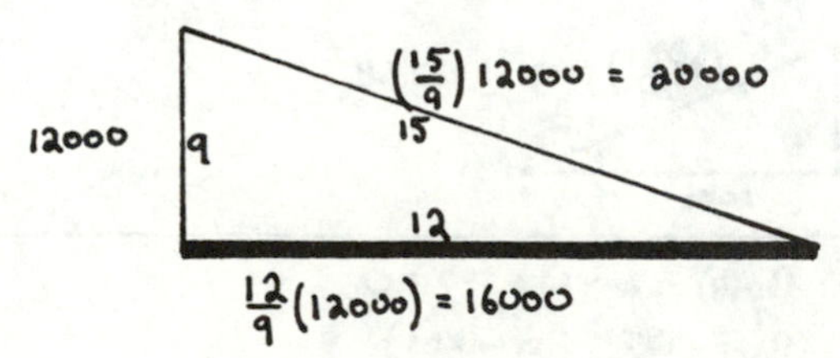

(MORE)

PROBLEM #14 CONTINUED

IN THE X-Z PLANE

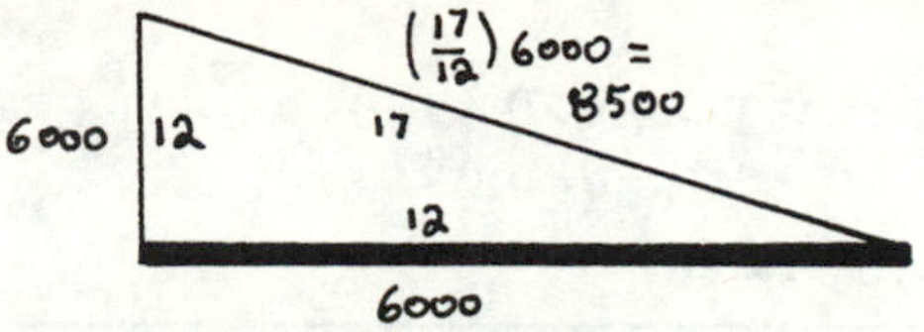

SO $A_x = 6000 + 16000 = 22000$

$A_y = 0$ $\quad A_z = 0$

$B_x = 16000$ $\quad B_y = 12000$ $\quad B_z = 0$

$C_x = 6000$ $\quad C_y = 0$ $\quad C_z = 6000$

15 REFER TO PAGES 9-9 AND 9-10. FROM FIGURE 9-12,

$w = 2$ LBM/FT

$S = 10$

$a = 50$

SOLVING EQN 9.51 BY TRIAL AND ERROR GIVES DISTANCE C:

$$S = c\left[\cosh\left(\frac{a}{c}\right) - 1\right]$$

$c = 126.6$

THE MIDPOINT {HORIZONTAL} TENSION IS GIVEN BY EQN 9.53:

$H = wc = (2)(126.6) = 253.2$ LBF

THE ENDPOINT {MAXIMUM} TENSION IS GIVEN BY EQN 9.55:

$T = wy = w(c+S)$

$= 2(126.6 + 10) = 273.2$

16 FROM EQN 9.55,

$T = wy$ OR

$y = \frac{T}{w} = \frac{500}{2} = 250$ AT RIGHT SUPPORT

FROM EQN 9.48,

$$250 = c\left(\cosh\left(\frac{50}{c}\right)\right)$$

$c = 245$ BY TRIAL AND ERROR

FROM EQN 9.51,

$$S = 245\left[\cosh\left(\frac{50}{245}\right) - 1\right] = 5.12 \text{ ft}$$

{OR, NOTICE FROM FIGURE 9.12 THAT $S = y - c = 5$}

17

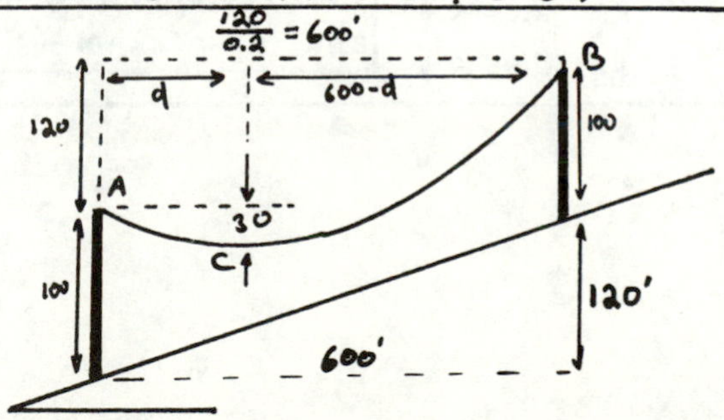

STEP 1 WORK FROM POINT A TO POINT C:

$S_{AC} = 30'$

FROM EQUATION 9.51,

$$30 = c\left[\cosh\left(\frac{d}{c}\right) - 1\right] \qquad \text{I}$$

STEP 2 WORK FROM POINT B TO C:

$S_{BC} = 150$

FROM EQUATION 9.51,

$$150 = c\left[\cosh\left(\frac{600-d}{c}\right) - 1\right] \qquad \text{II}$$

STEP 3 SOLVE EQUATION I FOR d:

$$d = c\left[\text{arccosh}\left(1 + \frac{30}{c}\right)\right]$$

STEP 4 SUBSTITUTE d INTO II:

$$150 = c\left[\cosh\left(\frac{600 - c\left[\text{arccosh}\left(1 + \frac{30}{c}\right)\right]}{c}\right) - 1\right]$$

SO $c = 591$ BY TRIAL AND ERROR.

STEP 5 $\quad d = 591\left[\text{arccosh}\left(1 + \frac{30}{591}\right)\right] = 187.5$ ft

FROM EQUATION 9.53,

$H = wc = (10)(591) = 5910$ LB

$T_A = (10)(30 + 591) = 6210$

$T_B = (10)(150 + 591) = 7410$

18 BY INSPECTION, $\bar{y} = 0$.

TO FIND $\bar{x}$, DIVIDE THE OBJECT INTO 3 PARTS:

$A_1 = 8(4) = 32$

$\bar{x}_1 = 2$

$A_2 = A_3 = 2(4) = 8$

$\bar{x}_2 = \bar{x}_3 = 6$

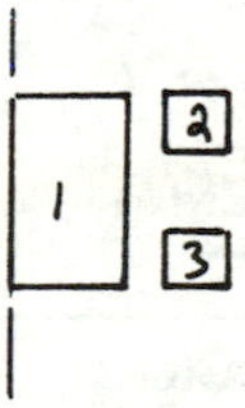

$$\bar{x}_c = \frac{32(2) + 8(6) + 8(6)}{32 + 8 + 8} = 3.333$$

19 DIVIDE INTO 2 PARTS:

$A_1 = (1)(8) = 8$

$\bar{y}_1 = -\frac{1}{2}$

$\bar{x}_1 = 4$

$A_2 = 1(5) = 5$

$\bar{x}_2 = \frac{1}{2}$

$\bar{y}_2 = -1 - 2.5 = -3.5$

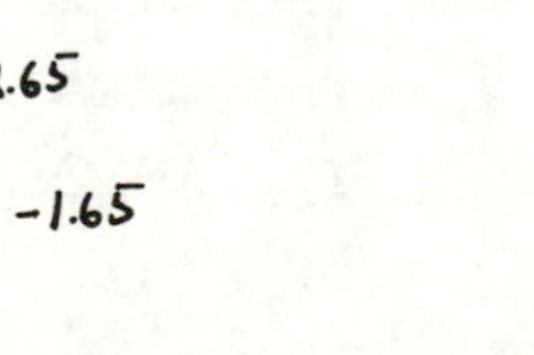

$$\bar{x}_c = \frac{(8)(4) + (5)\left(\frac{1}{2}\right)}{8+5} = 2.65$$

$$\bar{y}_c = \frac{(8)\left(-\frac{1}{2}\right) + (5)(-3.5)}{8+5} = -1.65$$

20 FOR THE PARABOLA

FROM PAGE 1-2

$A = \frac{2bh}{3} = \frac{2(300)(3)}{3}$

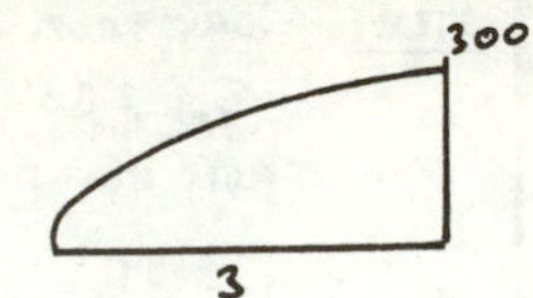

$= 600$ LBF

FROM PAGE 9-19 CENTROID IS

$\frac{3h}{5} = \frac{3(3)}{5} = 1.8'$ FROM TIP

DIVIDE THE REMAINING AREA INTO A TRIANGLE AND A RECTANGLE

RECTANGLE

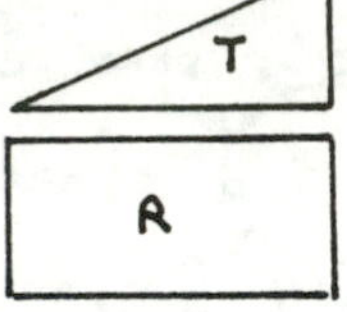

$A_R = (8)(300) = 2400$ LBF

CENTROID IS 4' FROM RIGHT END

TRIANGLE

$A_T = \frac{1}{2}(8)(700-300) = 1600$ LBF

CENTROID IS LOCATED

$\left(\frac{8}{3}\right) = 2.67$ FROM RIGHT END

21 $\bar{y} = 0$ BY INSPECTION

$\bar{x}_c = 0$

$\bar{x}_R = 1+3 = 4$

$\bar{x}_s = 1+6+1 = 8$

$$\bar{x}_{composite} = \frac{(64.4)(0)+(32.2)(4)+(64.4)(8)}{64.4+32.2+64.4}$$

$= 4$

22 $I_{A-A} = I_{C-C} + \frac{w}{g}(d_{C-A})^2$

OR $I_{C-C} = 90 - \frac{64.4}{32.2}(4)^2 = 58$

SO $I_{B-B} = 58 + \frac{64.4}{32.2}(6)^2 = 130$ SLUG-FT^2

23 DIVIDE INTO 3 AREAS

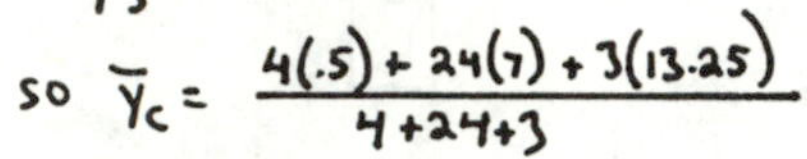

$A_1 = (4)(1) = 4$

$\bar{y}_1 = \frac{1}{2}$

$A_2 = (2)(12) = 24$

$\bar{y}_2 = 1+6 = 7$

$A_3 = (6)(\frac{1}{2}) = 3$

$\bar{y}_3 = 13.25$

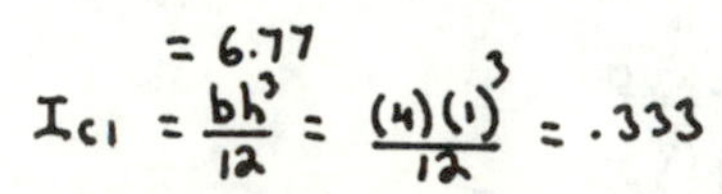

SO $\bar{y}_c = \frac{4(.5)+24(7)+3(13.25)}{4+24+3}$

$= 6.77$

$I_{c1} = \frac{bh^3}{12} = \frac{(4)(1)^3}{12} = .333$

$d_1 = 6.77 - .5 = 6.27$

$I_{c2} = \frac{(2)(12)^3}{12} = 288$

$d_2 = .23$

$I_{c3} = \frac{(6)(\frac{1}{2})^3}{12} = .0625$

$d_3 = 6.48$

USING THE PARALLEL AXIS THEOREM (EQN 9.80)

$I_{total} = .333 + 4(6.27)^2 + 288 + 24(.23)^2 + .0625 + 3(6.48)^2$

$= 572.88$

24 DIVIDE THE AREA INTO 3 PARTS

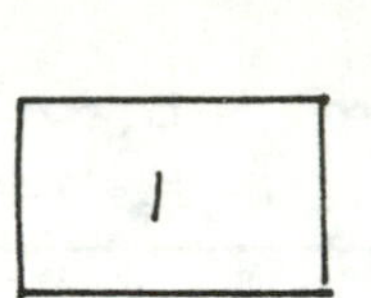

$A_1 = (4)(6) = 24$

$\bar{y}_1 = 2$

$I_{c1} = \frac{6(4)^3}{12} = 32$

$A_2 = \frac{1}{2}(6)(4) = 12$

$\bar{y}_2 = 4 + \frac{4}{3} = 5.333$

$I_{c2} = \frac{1}{36}(6)(4)^3 = 10.67$

$A_3 = \frac{1}{2}(\pi(2)^2) = -6.283$

$\bar{y}_3 = \frac{4(2)}{3\pi} = .849$

$I_{c3} = .11r^4 = .11(2)^4 = -1.76$

$$\bar{y}_{composite} = \frac{24(2)+12(5.333)-6.283(.849)}{24+12-6.283}$$

$= 3.59$

$d_1 = 3.59 - 2 = 1.59$

$d_2 = 5.333 - 3.59 = 1.743$

$d_3 = 3.59 - .849 = 2.741$

$I_{total} = 32 + 24(1.59)^2 + 10.67 + 12(1.743)^2 - 1.76 - 6.283(2.741)^2$

$= 90.84$

25 $K = \sqrt{I/A}$

$\sqrt{\frac{90.84}{24+12-6.283}} = 1.748$

26 $R_1 = 13.41$ K ↑

$R_2 = 2.67$ K ↑

$R_3 = 2.48$ K ↓

A PROBLEM OF THIS TYPE WOULD NOT BE ON THE E-I-T EXAM

E-I-T HOMEWORK SOLUTIONS: DYNAMICS

1

$x(t) = 2t^2 - 8t + 3$

$v(t) = \frac{dx}{dt} = 4t - 8$

$a(t) = \frac{dv}{dt} = 4$

AT $t = 2$

$x = 2(2)^2 - 8(2) + 3 = -5$

$v = 4(2) - 8 = 0$

$a = 4$

AT $t = 1$

$x = 2(1) - 8(1) + 3 = -3$

AT $t = 3$

$x = 2(3)^2 - 8(3) + 3 = -3$

DISPLACEMENT $= X_3 - X_1 = -3 - (-3) = 0$

USE A TABULAR METHOD TO FIND TOTAL DISTANCE TRAVELED.

$x(1) = -3$

$x(2) = -5$ AND $v(2) = 0$

$x(3) = -3$

SO, THE PARTICLE SLOWS DOWN BETWEEN $t=1$ AND $t=2$, STOPS, AND SPEEDS UP (IN THE OPPOSITE DIRECTION) BETWEEN $t=2$ AND $t=3$.

DISTANCE $= (-3-(-5)) + (-3-(-5)) = 4$

2

$\omega(2) = 6(2)^2 - 10(2) = 4$ CW

$\theta(t) = \int \omega(t) = 2t^3 - 5t^2$

$\theta(1) = 2(1)^3 - 5(1)^2 = -3$

$\theta(3) = 2(3)^3 - 5(3)^2 = 9$

$\Delta\theta = 9 - (-3) = 12$

TO FIND THE TOTAL DISTANCE TRAVELED, CHECK TO SEE IF THERE IS ANY SIGN REVERSAL IN $\omega(t)$ OVER THE INTERVAL $t=1$ TO $t=3$

$\omega(1) = -4$

$\omega(3) = 24$

SO, THE DIRECTION OF TRAVEL DOES CHANGE, AND THE TOTAL DISTANCE TRAVELED IS NOT THE SAME AS THE DISPLACEMENT.

FIND THE TIME t FOR $\omega = 0$

$6t^2 - 10t = 0$

$6t - 10 = 0$

$t = 10/6$

$\theta\left(\frac{10}{6}\right) = 2\left(\frac{10}{6}\right)^3 - 5\left(\frac{10}{6}\right)^2 = -4.63$

TOTAL DISTANCE TRAVELED

$(-3-(-4.63)) + (9-(-4.63)) = 15.26$

3

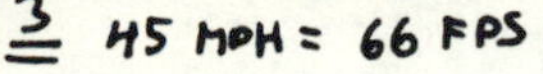

45 MPH = 66 FPS

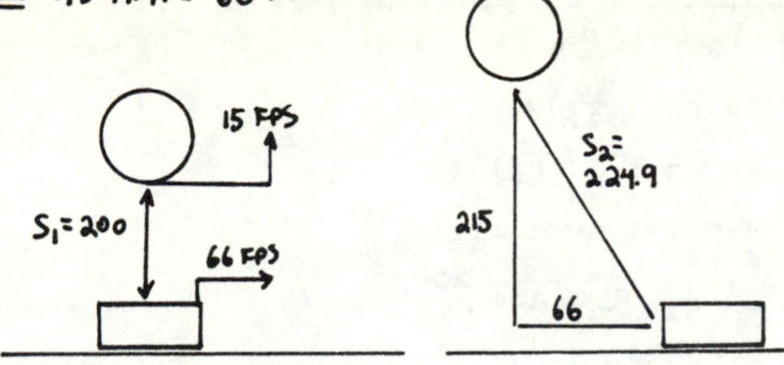

AT $t=0$, SEPARATION $= 200$ FT

AT $t=1$, $S_2 = \sqrt{(215)^2 + (66)^2} = 224.9$

THE SEPARATION VELOCITY IS THE DIFFERENCE IN VELOCITIES ALONG A MUTUALLY PARALLEL LINE.

SO, BREAK THE BALLOON'S AND CAR'S VELOCITIES INTO COMPONENTS PERPENDICULAR AND PARALLEL TO THE LINE S_2

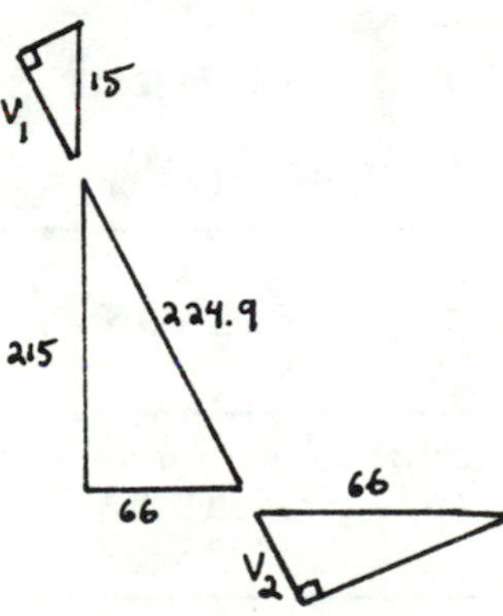

$V_1 = 15\left(\frac{215}{224.9}\right) = 14.34$

$V_2 = 66\left(\frac{66}{224.9}\right) = 19.37$

$\Delta V = 19.37 + 14.34 = 33.71$ FPS

4 FREEBODY OF A

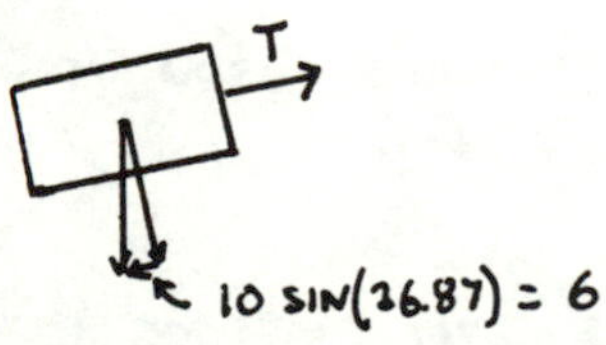

SO, THE NET ACCELERATION FORCE PARALLEL TO THE PLANE IS

$F_A = T - 6$

FROM $F = ma$,

$T - 6 = \left(\frac{10}{32.2}\right)a_A$

$T = 6 + \left(\frac{10}{32.2}\right)a_A$

FREEBODY OF B

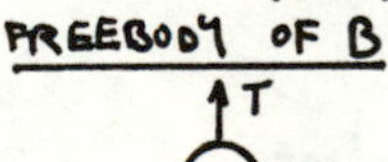

(MORE)

PROBLEM #4 CONTINUED

$F_B = M_B a_B$

$(20-T) = \left(\frac{20}{32.2}\right) a_B$

$T = 20 - \left(\frac{20}{32.2}\right) a_B$

BUT T=T AND $a_A = a_B$

$6 + \frac{10}{32.2} a = 20 - \frac{20}{32.2} a$

$a = 15.03$ FT/sec²

$V = at = (15.03)(3) = 45.1$ FT/SEC

5 $a = \frac{\Delta V}{\Delta t} = \frac{(180)\ MI/HR\ (5280)\ FT/MI}{(60)\ sec\ (3600)\ sec/HR} = 4.4$ FT/SEC²

6 $a = \frac{\Delta V}{\Delta t} = \frac{(20-5)\ FT/sec}{(2)\ MIN\ (60)\ \frac{sec}{MIN}} = .125$ FT/sec²

7 60 MPH = 88 FT/SEC

FROM TABLE 12.1

$a = \frac{88}{5} = 17.6$ FT/sec²

$S = \frac{1}{2} t V_o = (.5)(5)(88) = 220$ FT

8 $F = Ma$

$a = \frac{F}{M} = \frac{12}{7.8} = 1.54$ m/sec²

9 CIRCUMFERENCE $= \pi d = \pi(14) = 43.98$ IN

$V = (43.98)\ IN \left(\frac{40\ RPM}{60\ \frac{sec}{MIN}}\right) = 29.32$ IN/sec

10 $\omega = 2\pi f$

$\omega_1 = 2\pi\left(\frac{1200}{60}\right) = 125.66$ RAD/sec

$\omega_2 = 2\pi\left(\frac{3000}{60}\right) = 314.16$ RAD/sec

$\alpha = \frac{\Delta\omega}{t} = \frac{314.16 - 125.66}{10} = 18.85$ RAD/sec²

11 SPEED OF ROTATION:

$\frac{(1750)\ RPM\ (360)\ °/REV}{(60)\ \left(\frac{sec}{MIN}\right)} = 10500$ °/sec

TIME BETWEEN DISCS

$= \frac{18°}{10500\ °/sec} = 1.714$ EE-3 sec

$V = \frac{X}{t} = \frac{5}{1.714\ EE\text{-}3} = 2916.6$ FPS

12 $\omega_B = \left(\frac{16}{24}\right)\omega_C$

$= (.667)(2) = 1.333$ RAD/sec

SIMILARLY,

$\alpha_B = (.667)(6) = 4$ RAD/sec

BUT $\omega_A = \omega_B$ AND $\alpha_A = \alpha_B$

SO $V_D = \omega r_A = (1.333)(6) = 8$ IN/sec

$a_D = \alpha r_A = (4)(6) = 24$ IN/sec²

13 60 MPH = 88 FPS

AS IN EXAMPLE 12.5

$\tan\theta = \frac{V^2}{gr} = \frac{(88)^2}{(32.2)(6000)} = .04$

14 THE OBJECT IS ACTED UPON BY TENSION, GRAVITY, AND CENTRIFUGAL FORCE

F_c, T, 8.05

$F_c = \frac{MV^2}{r} = \frac{(8.05)}{32.2}\left(\frac{(20)^2}{4}\right) = 25$

THE CORD TENSION IS THEN

25, 8.05, θ

$\theta = \arctan\left(\frac{25}{8.05}\right) = 72.15°$

15 $r = \frac{50\ CM}{100\ \frac{CM}{M}} = .5$ M

$V = 2\pi r(RPS) = 2\pi(.5)(5) = 15.708$ m/sec

a) $a_N = \frac{V^2}{r} = \frac{(15.708)^2}{.5} = 493.5$ m/sec²

b) $F = Ma_N = (5)\ KG\ (493.5)\ M/sec^2$
$= 2467.5$ N {OUTWARD}

c) -2467.5 N {INWARD}

d) $I\omega = (Mr^2)\omega$
$= (5)\ KG\ (.5)^2\ m^2 (5) \frac{REV}{SEC} (2\pi) \frac{RAD}{REV}$
$= 39.27\ \frac{KG\text{-}M^2}{SEC}$

16 FROM EQN 12.37

$h = \frac{V_o^2 \sin^2\phi}{2g} = \frac{(2700)^2(\sin 45)^2}{2(32.2)} = 56,599$ FT

$r = \frac{V_o^2 \sin 2\phi}{g} = \frac{(2700)^2 \sin 90°}{32.2} = 226,398$ FT

17 FROM EQN 12.35

$V_x = 60(\cos 36.87°) = 48$ {CONSTANT}

$t = \frac{72}{48} = 1.5$ sec

FROM EQN 12.36

$V_y = 60(\sin 36.87) - 32.2(1.5) = -12.3$

FROM EQN 12.33

$y = (60)(\sin 36.87)(1.5) - \frac{1}{2}(32.2)(1.5)^2$
$= 17.78$ FT

18 $h_{MAX} = 12000 + \frac{(600)^2 \sin^2 30°}{2(32.2)}$

$= 12000 + 1398 = 13398$ FT

{MORE}

PROBLEM #18 CONTINUED

$t_{\text{TO MAX HT}} = \frac{1}{2}\left(\frac{2V_0 \sin\phi}{g}\right) = \frac{(600)\sin 30°}{32.2} = 9.317 \text{ SEC}$

$t_{\text{TO FALL 13398 FT}} = \sqrt{\frac{2h}{g}} = \sqrt{\frac{(2)(13398)}{32.2}} = 28.847$

$t_{\text{total}} = 38.16$ SEC

19

d = 1000 FT

$F = \frac{1}{2}(1000)\text{ FT}(2)\text{ LBM/FT} = 1000$ LBM

$W = Fd = (1000)(1000) = $ EE6 FT-LBF

20 FROM TABLE 11.1

$\rho_{\text{CAST IRON}} = 442 \text{ PCF} = .256 \text{ LBM/IN}^3$

$\text{VOL} = \frac{4}{3}\pi r^3 = \frac{4}{3}\pi\left(\frac{10}{2}\right)^3 = 523.6$

$M = \frac{(\text{VOL})\rho}{32.2} = \frac{(523.6)(.256)}{32.2} = 4.163$ SLUGS

$E_K = \frac{1}{2}MV^2 = (.5)(4.163)(30)^2$

$= 1873$ FT-LBF

21 ENERGY AVAILABLE

$= (5)\text{ KW}(1000)\frac{\text{W}}{\text{KW}}(.85)(1)\frac{\text{JOULE}}{\text{SEC-WATT}}(1)\text{ HR}(3600)\frac{\text{SEC}}{\text{HR}}$

$= 1.53$ EE7 JOULES

1 M^3 OF WATER HAS A MASS OF 1000 KG

$\frac{(1.53 \text{ EE7})\text{ J}}{(1000)\text{ KG/M}^3\ (9.81)\text{ M/SEC}^2\ (40)\text{ M}} = 39 \text{ M}^3$

22 ASSUME g IS CONSTANT

$W = \Delta E_P$

$= Mgh = (5.2)\text{ KG}(9.81)\text{ M/SEC}^2(12000)\text{ M}$

$= 6.12$ EE5 J

$P = \frac{W}{\Delta t} = \frac{6.12\text{EE5 J}}{(30)\text{MIN}(60)\text{SEC/MIN}} = 340$ W

23 HEIGHT $= (8)\text{ FT}(12)\text{ IN/FT} = 96$ IN

POTENTIAL ENERGY ABSORBED

$= (100)\text{ LBM}(96+\delta)\text{ IN}$

$W_{\text{DONE}} = \frac{1}{2}k\delta^2 = (\frac{1}{2})(33.33)\delta^2$

SO $(100)(96+\delta) = \frac{1}{2}(33.33)\delta^2$

OR $\delta = 27.19$ IN

$= 2.266$ FT

24 $W = \Delta E_K$

$\Delta E_K = \frac{1}{2}I\Delta(\omega)^2$

$\omega = 2\pi\left(\frac{\text{RPM}}{60}\right)$

$4500 = \frac{1}{2}(15)\left(\frac{2\pi}{60}\right)^2\left[(300)^2 - (\text{RPM}_2)^2\right]$

$(\text{RPM}) = 187.8$

25 IGNORE ANY TIPPING TENDENCIES

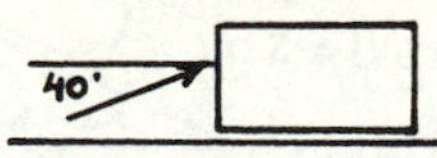

$F_x = (2500)\cos 40 = 1915.11$

$W = (F_x)d = (1915.11)(6) = 11490.6$ J

26 $W = (1500)\text{KG}(9.81)\text{ M/SEC}^2(80)\text{ M} = 1177200$ J

$P = \frac{W}{t} = \frac{(1177200)\text{ J}(1)\text{ W-SEC/J}}{(14)\text{ SEC}(1000)\text{ W/KW}}$

$= 84.09$ KW

27 INITIAL KINETIC ENERGY

$E_K = \frac{1}{2}MV^2 = \frac{1}{2}\left(\frac{100}{32.2}\right)(12.88)^2 = 257.6$ FT-LBF

N = 100 LBF

$F_f = fN = .2(100) = 20$

$W = (F_f)d$

$257.6 - 20(d) = 0$

$d = 12.88$ FT

28 THE FRICTIONAL FORCE IS

$F_f = (.333)\,W$

THIS IS ALSO THE ACCELERATING FORCE

$F = Ma$

$(.333)W = \left(\frac{W}{g}\right)a$

$a = (.333)(32.2) = 10.72$

$t = \frac{v}{a} = \left(\frac{10}{10.72}\right) = .9328$ SEC

29 40 MPH = 58.67 FPS

THE CENTRIFUGAL FORCE IS

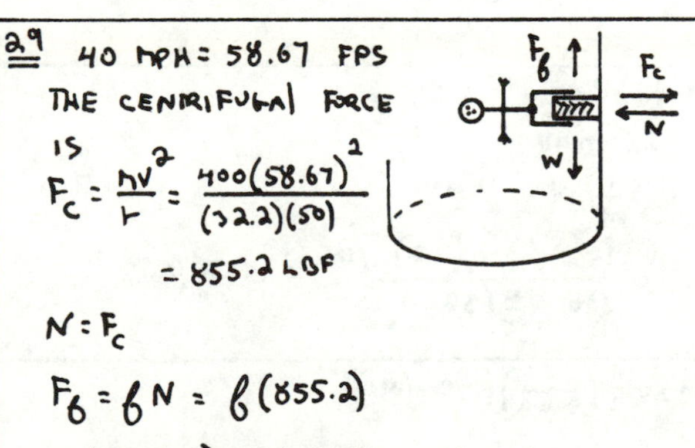

$F_c = \frac{MV^2}{r} = \frac{400(58.67)^2}{(32.2)(50)}$

$= 855.2$ LBF

$N = F_c$

$F_f = fN = f(855.2)$

$f(855.2) = W = 400$

$f = .468$

30 FROM TABLE 12.1

$a = \frac{(12)^2}{2(36)} = 2 \text{ FT/SEC}^2$

$F_f = (.25)(10) = 2.5$

$F = Ma$

$P - 2.5 = \left(\frac{10}{32.2}\right)(2)$

$P = 3.12$ LBF

31

$N = 130\cos(22.64) = 120$

$W_x = 130\sin(22.64) = 50$

$F_f = fN = (.1)(120) = 12$

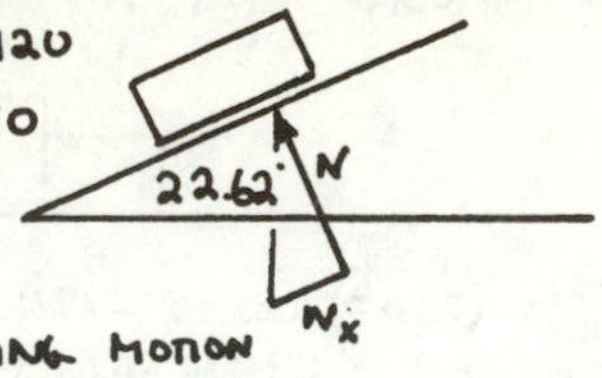

SO, THE FORCES RESISTING MOTION ARE $50 + 12 = 62$

$F = ma$

$62 = \left(\frac{130}{32.2}\right)a$

$a = 15.36\ \text{FT/SEC}^2$

FROM TABLE 12.1,

$$S = \frac{(30)^2 - (0)^2}{2(15.36)} = 29.3\ \text{FT}$$

32

THE FRICTIONAL FORCE F_{f1} IS

$F_{f1} = fN$

$= .2(100+50)$

$= 30$ LBF

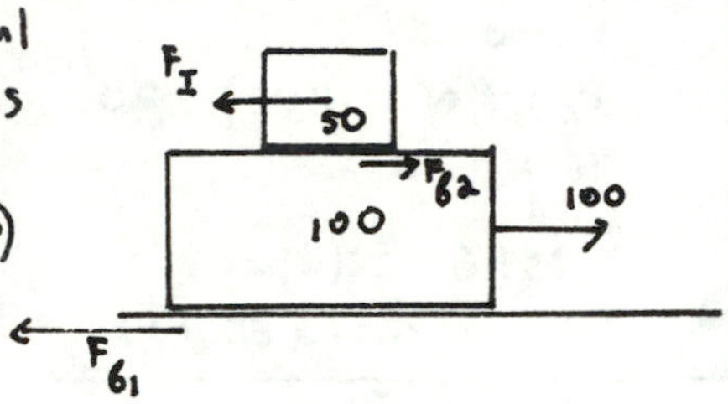

SO, THE ACCELERATION OF THE BLOCK COMBINATION IS $a = \frac{F}{m} = \frac{100-30}{\left(\frac{150}{32.2}\right)} = 15.03\ \text{FT/SEC}^2$

THE INERTIAL RESISTING FORCE F_I IS

$F_I = ma = \left(\frac{50}{32.2}\right)(15.03) = 23.34$ LBF

IF THIS IS RESISTED BY FRICTION,

$23.34 = f(50)$

$f = .467$

33 FROM EQN 12.59,

$F\Delta t = m\Delta V$

$F = \frac{m}{\Delta t}\Delta V = \dot{m}\Delta V$

$$= \frac{(560)\ \frac{\text{LBM}}{\text{MIN}}\ (3.2)\ \text{FT/SEC}}{(60)\ \frac{\text{SEC}}{\text{MIN}}\ (32.2)\ \frac{\text{FT}}{\text{SEC}^2}} = .928\ \text{LBF}$$

34 $m\Delta V = \left(\frac{.4}{32.2}\right)(130-(-90))$

$= 2.73$ LBF-SEC

35 $m_1V_1 = m_2V_2$

$(500)V_1 = (1.2)(650)$

$V_1 = 1.56$ M/SEC

36 $m_1V_1 = (m_1+m_2)V_2$

$(.06)(700) = (.06+4.5)V_2$

$V_2 = 9.21$ M/SEC

37 $(40)\ \text{gpm}\ (2.228\ \text{EE-3})\ \frac{\text{FT}^3}{\text{SEC-gpm}}$

$= 8.912$ EE-2 FT^3/SEC

$$\dot{m} = \frac{(8.912\ \text{EE-2})\ \text{FT}^3/\text{SEC}\ (62.4)\ \text{LBM/FT}^3}{(32.2)\ \text{FT/SEC}^2}$$

$= 0.1727$ SLUG/SEC

$\Delta V = 60$ FPS

$F\Delta t = m\Delta V$

$F = \dot{m}\Delta V = (.1727)(60) = 10.36$ LBF

38 $(100)\ \text{gps}(.1337)\ \frac{\text{FT}^3}{\text{gal}} = 13.37\ \text{FT}^3/\text{SEC}$

$\dot{m} = \frac{(13.37)(62.4)}{32.2} = 25.91$ SLUG/SEC

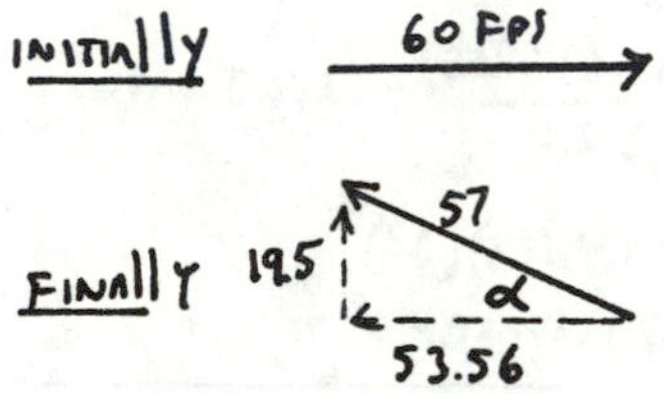

$\alpha = 180° - 160° = 20°$

$F_x = \dot{m}\Delta V_x = (25.91)(60+53.56) = -2942.3$ LBF

$F_y = \dot{m}\Delta V_y = (25.91)(19.5) = +505.2$

$R = \sqrt{F_x^2 + F_y^2} = 2985.4$

$\theta = \arcsin\left(\frac{505.2}{-2985.4}\right) = -9.74°\ (170.3°)$

39 FROM PAGE 1-38,

$m_{electron} = 9.11$ EE-31 KG

$m_{hydrogen} = m_{proton} = 1.673$ EE-27 KG

THE ORIGINAL MOMENTUM IN THE X DIRECTION IS

$P_1 = (9.11\ \text{EE-31})(500) = (4.555\ \text{EE-28})\ \frac{\text{KG-M}}{\text{SEC}}$

AFTER THE COLLISION,

$P_{electron} = (9.11\ \text{EE-31})(490)\cos 30°$

$= 3.866$ EE-28

$P_{hydrogen} = (1.673\ \text{EE-27})\,V_{xH}$

$P_1 = P_2$

$(4.555\ \text{EE-28}) = (3.866\ \text{EE-28}) + (1.673\ \text{EE-27})V_{Hx}$

$V_{Hx} = 4.088$ EE-2 M/SEC

NOTICE THAT THE QUESTION ASKED FOR V_{Hx}, NOT V_H

40 $V_L = 5 \quad V_R = 4$

FROM EQN 12.63,

$5(5) + 5(-4) = (5+5)V_2$

$V_2 = .5$ FPS

41

$V_{Ay} = 5\left(\frac{1}{\sqrt{2}}\right)$ {CONSTANT}

$V_{Ax} = 5\left(\frac{1}{\sqrt{2}}\right)$

$V_{By} = 10\left(\frac{3}{5}\right) = 6$ {CONSTANT}

$V_{Bx} = 10\left(\frac{4}{5}\right) = -8$

FROM EQN 12.61,

$$.8 = \frac{-(V'_{Bx} - V'_{Ax})}{-8 - \frac{5}{\sqrt{2}}}$$

$$9.228 = V'_{Bx} - V'_{Ax}$$

FROM EQN 12.63,

$$2\left(\frac{5}{\sqrt{2}}\right) + 2(-8) = 2(V'_{Ax}) + 2(V'_{Bx})$$

OR $-4.464 = V'_{Ax} + V'_{Bx}$

SOLVING SIMULTANEOUSLY,

$$V'_{Ax} = -6.846$$

$$V'_{Bx} = 2.382$$

$$V'_A = \sqrt{\left(\frac{5}{\sqrt{2}}\right)^2 + (-6.846)^2} = 7.705$$

$$\phi_A = \text{ARCTAN}\left(\frac{-5}{\sqrt{2}(6.846)}\right) = 152.7°$$

$$V'_B = \sqrt{(6)^2 + (2.382)^2} = 6.456$$

$$\phi_B = \text{ARCTAN}\left(\frac{6}{2.382}\right) = 68.35$$

42

a) $mgh = \frac{1}{2}mv^2$

$$v = \sqrt{2gh} = \sqrt{(2)(32.2)(3)} = 13.9 \text{ FPS}$$

b) WT = 10 LBM

$$F_c = \frac{mv^2}{r} = \frac{(10)(13.9)^2}{(32.2)(3)} = 20$$

TENSION = 10 + 20 = 30

c) FROM EQN 12.61,

$$.7 = \frac{-(V'_2 - V'_1)}{0 - 13.9}$$

$$9.73 = V'_2 - V'_1$$

FROM EQN 12.63,

$$10(13.9) + 50(0) = 10(V'_1) + 50(V'_2)$$

$$139 = 10V'_1 + 50V'_2$$

SOLVING SIMULTANEOUSLY,

$$V'_1 = -5.79$$

$$V'_2 = 3.94$$

d) $\frac{1}{2}mv^2 = \frac{1}{2}kx^2$

$$\left(\frac{50}{32.2}\right)(3.94)^2 = k(6/12)^2$$

$$k = 96.4 \text{ LB/FT}$$

$$= 8.03 \text{ LB/IN}$$

43 CHOOSE AXES PARALLEL AND PERPENDICULAR TO PLANE.

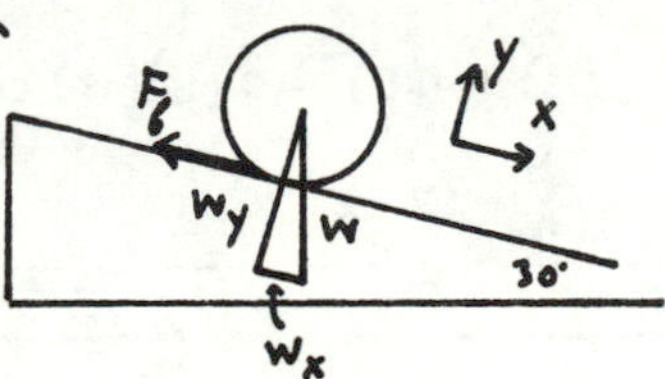

FOR THE LINEAR TRANSLATION

$$W_x = W \sin 30°$$

FROM $F = ma$

$$W(\sin 30°) - F_f = \left(\frac{W}{g}\right)a \quad \text{(I)}$$

FOR THE ROTATION

$$\Sigma T = I\alpha$$

$$(F_f)r = \left(\frac{2}{5}mr^2\right)\alpha$$

BUT $a = r\alpha$, SO

$$F_f = \left(\frac{2}{5}\right)ma \quad \text{(II)}$$

COMBINING (I) AND (II),

$$W\left(\frac{1}{2}\right) - \left(\frac{2}{5}\right)ma = ma$$

$$\frac{g}{2} = \left(\frac{7}{5}\right)a$$

$$a = 11.5 \text{ FT/SEC}^2$$

$$V = at = 2(a) = 2(11.5) = 23 \text{ FT/SEC}$$

44

a) $F = ma$

$$20 = \left(\frac{100}{32.2}\right)a$$

SO

$$a = 6.44 \text{ FT/SEC}^2$$

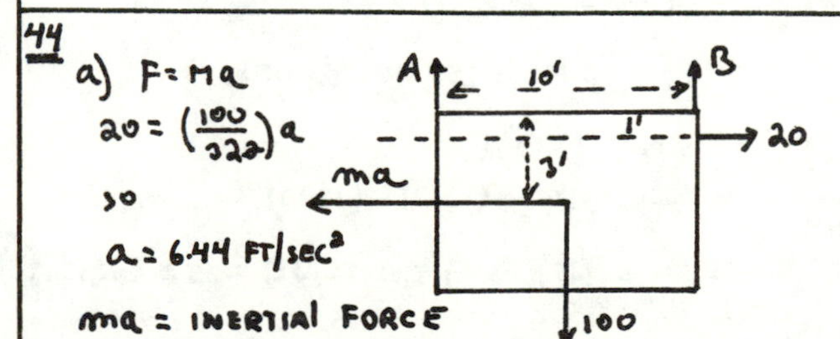

ma = INERTIAL FORCE

b) ΣM_A:

$$10B + 20(1) - 20(3) - 100(5) = 0$$

$$B = 54$$

$$A = 100 - 54 = 46$$

c) The FORCE would HAVE TO BE APPLIED IN LINE WITH THE CENTER OF GRAVITY – OPPOSITE THE MA INERTIAL FORCE.

45

a) $V_{A-O} = \left(\frac{3}{2}\right)10 = 15$ ↘

b) $V_{B-O} = 15$ ←

46 ADD THE NEGATIVE OF A'S VELOCITY TO B'S VELOCITY:

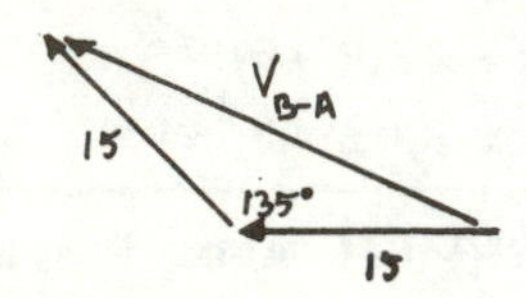

$V_{B-A}^2 = (15)^2 + (15)^2 - 2(15)(15)\cos 135°$

OR $V_{B-A} = 27.72$

47

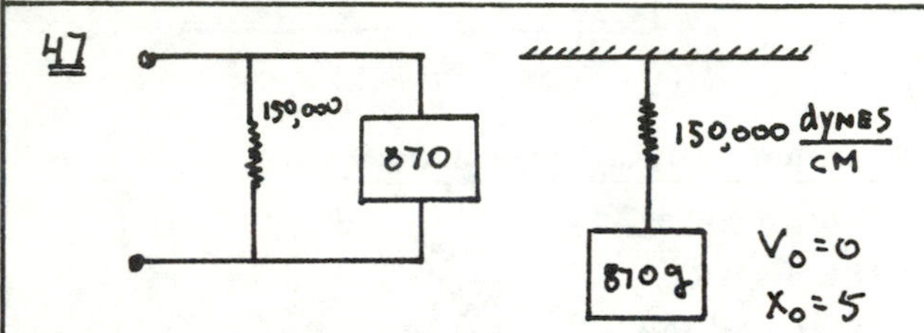

THE EQUATION OF MOTION IS

$0 = 150{,}000\,x + 870\,x''$

OR, $0 = 172.4\,x + x''$

TAKING THE LAPLACE TRANSFORM,

$s^2\mathcal{L}(x) - s(5) + 172.4\,\mathcal{L}(x) = 0$

$\mathcal{L}(x)\left[s^2 + 172.4\right] = 5s$

$\mathcal{L}(x) = 5\left[\frac{s}{s^2+172.4}\right]$

$x = 5\cos(13.13\,t)$

$\omega = 13.13$ RAD/SEC

$f = \frac{13.13}{2\pi} = 2.09$ Hz

b) $T = 1/f = 1/2.09 = .478$

c) $F = 5(150{,}000) = 750{,}000$ dynes

d) $x = 5\cos(13.13)t$

$v = \frac{dx}{dt} = -5\sin(13.13)t\,(13.13)$

$a = \frac{dv}{dt} = (-5)(13.13)^2\cos(13.13t) = 862\cos(13.13t)$

$a_{max} = 862$

e) WHEN $x = 3$,

$3 = 5\cos(13.13\,t)$

OR $t = .07062$ SEC

$v = -(5)(13.13)\sin\left[13.13(.07062)\right]$

$= -52.5$ cm/sec {SIGN IRRELEVANT}

f) V IS MAX AT $t = \frac{\pi}{4}$ OR $\theta = \pi/2$

$v = (-5)(13.13)\sin\left(\frac{\pi}{2}\right) = 65.65$

48

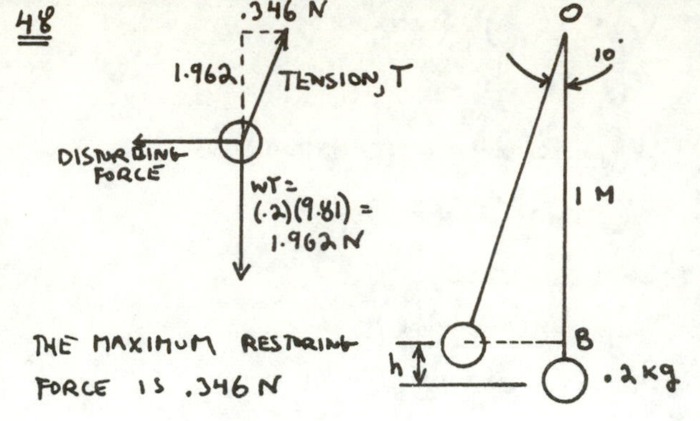

THE MAXIMUM RESTORING FORCE IS .346 N

b) $F = ma$

$a = \frac{F}{m} = \frac{.346\text{ N}}{.2\text{ kg}} = 1.73$ m/sec²

$\alpha = \frac{a}{r} = \frac{1.73}{1} = 1.73$

c) $OB = 1\cos 10°$; $h = 1 - OB = .0152$

$v = \sqrt{2gh} = \sqrt{(2)(9.81)(.0152)}$

$= .55$ m/sec

49 $J = mr^2 = (24)g(.8)^2$

$= 15.36$ g-cm²

b) $T = 2\pi\sqrt{J/K}$

$.2 = 2\pi\sqrt{15.36/K}$

$\left(\frac{.2}{2\pi}\right)^2 = \frac{15.36}{K}$

$K = 15160$ $\frac{\text{DYNE-CM}}{\text{RADIAN}}$

50 ROTATIONAL MOMENT OF INERTIA OF A DISK THROUGH ITS CENTER IS $\frac{1}{2}mr^2$, ABOUT A POINT ON ITS RIM,

$I = \frac{1}{2}mr^2 + mr^2 = \frac{3}{2}mr^2$

THEN, $T = 2\pi\sqrt{I/mgh} = 2\pi\sqrt{3r/2g}$

51

$F_{gravity} = -(2\delta)\rho A$

$M_{motion} = \rho AL/g$

$a = d^2\delta/dt^2 = \delta''$

$F = ma$

$-2\delta\rho A = \frac{\rho AL}{g_c}\delta''$

$\delta'' + \left(\frac{2g_c}{L}\right)\delta = 0$

THIS HAS A SOLUTION OF $f = \frac{1}{2\pi}\sqrt{\frac{2g_c}{L}}$

$f = \frac{1}{2\pi}\sqrt{\frac{(2)(32.2)(12)}{30}} = .81$ Hz

E-I-T HOMEWORK SOLUTIONS: MECHANICS OF MATERIALS

1 AS LIMITED BY STRESS

$\sigma = \frac{F}{A}$ OR

$A = \frac{F}{\sigma} = \frac{30,000}{10,000} = 3\ IN^2$

AS LIMITED BY ELONGATION

$\delta = \frac{FL}{AE}$ OR

$A = \frac{FL}{\delta E} = \frac{(30,000)\ LBF\ (200)\ IN}{(.02)\ IN\ (3\ EE7)\ PSI} = 10\ IN^2$

SO, $10\ IN^2$ IS THE REQUIRED AREA.

2 a) $\delta = L\epsilon$

$= (16)\ IN\ (.0012) = .0192''$

b) $\sigma = E\epsilon$

$= (2.5\ EE6)(.0012) = 3000\ PSI$

3 $\sigma_{MAX} = \frac{60,000}{5} = 12000$

$\sigma = \frac{F}{A} = \frac{4F}{\pi d^2}$

$d = \sqrt{\frac{4F}{\pi\sigma}} = \sqrt{\frac{(4)(7000)}{(\pi)(12000)}} = .8618''$

4 THE MAXIMUM AREA IN SHEAR $= \pi\left(\frac{3}{4}\right)\left(\frac{5}{8}\right) = 1.4726\ IN^2$

$F = \sigma A = (42,000)(1.4726) = 61850\ LBF$

5 $F_\ell = 40,000$

$\sigma_D = \mu\sigma_\ell$

$\delta_D = \frac{F_D D}{AE} = \frac{\mu F_\ell D}{AE}$

$= \frac{(.3)(40,000)\left(\frac{3}{4}\right)}{\frac{\pi}{4}\left(\frac{3}{4}\right)^2 3\ EE7} = 6.791\ EE\text{-}4$

SO THE HOLE DIAMETER IS

$D + \delta = .750679$

6 $E = \frac{\Delta\sigma}{\Delta\epsilon} = \frac{\Delta F/A}{\Delta L/L_o}$

$= \frac{1500}{\frac{\pi}{4}\left(\frac{5}{16}\right)^2} \Big/ \frac{6}{1500(12)} = 58.7\ EE6\ PSI$

7 $\Delta\sigma = E\epsilon_{th} = E\alpha\Delta T$

$= (3\ EE7)(6.5\ EE\text{-}6)(90) = 17550$

SO, $\sigma = 17550 - 2000 = 15550\ PSI\ (T)$

8 $\epsilon_{th} = \alpha\Delta T = (6.5\ EE\text{-}6)(80) = 5.2\ EE\text{-}4$

$\sigma = E\epsilon = (3\ EE7)(5.2\ EE\text{-}4) = 1.56\ EE4\ PSI$

$F = \sigma A = (1.56\ EE4)\left(\frac{1}{2}\right)(4) = 31200\ LBF$

9 ERROR DUE TO TENSION

$\delta = \frac{\Delta F L_o}{AE} = \frac{(10)(100)(12)}{\left(\frac{1}{32}\right)\left(\frac{3}{8}\right)(3\ EE7)} = .03413''$

ERROR DUE TO TEMPERATURE

$\delta = \epsilon L_o = (6.5\ EE\text{-}6)(30)(100)(12)$

$= .234$

$\delta_{total} = \frac{.234 + .03413}{12} = .0223\ FT$

$L_{ACTUAL} = 100.0223\ FT$

10

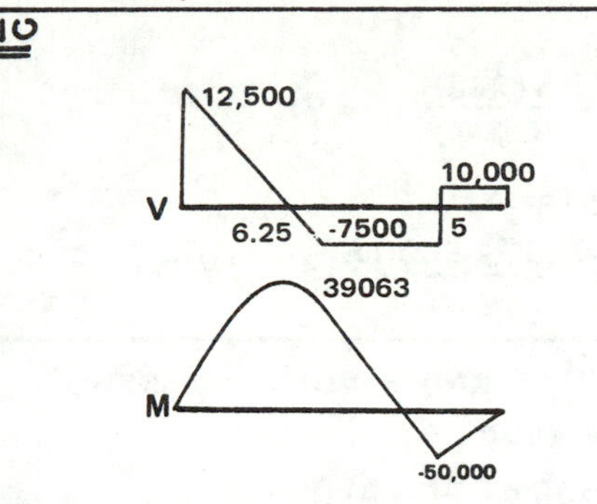

11

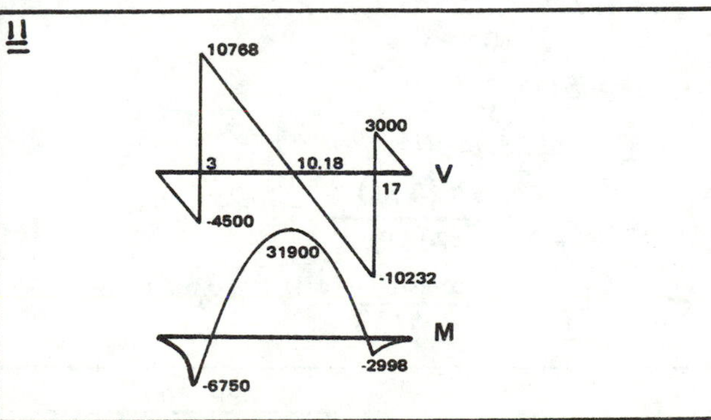

12 $M_{max} = 20058$ AT $X = 12$ FROM LEFT END.

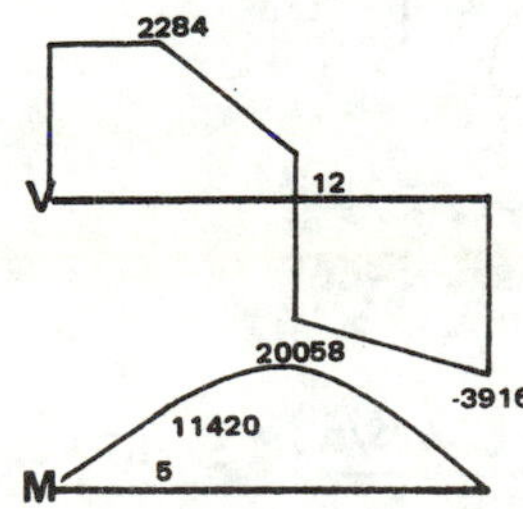

13 $M_{max} = 15000$ AT $x = 3$ FROM LEFT END.

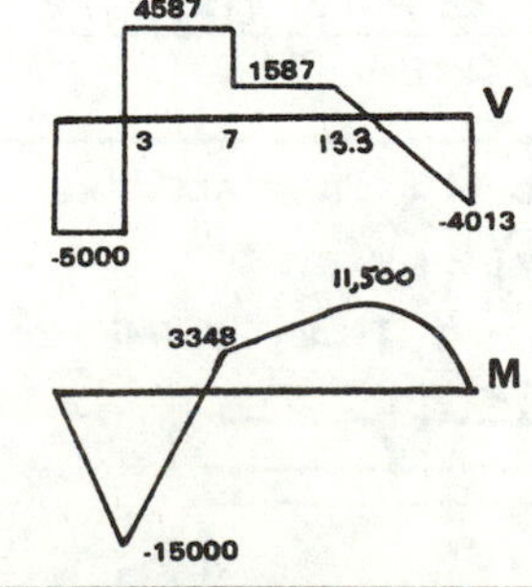

14 $A = 10(20) = 200\ IN^2$

$I = \frac{10(20)^3}{12} = 6666.7\ IN^4$

(MORE)

PROBLEM 14, CONTINUED

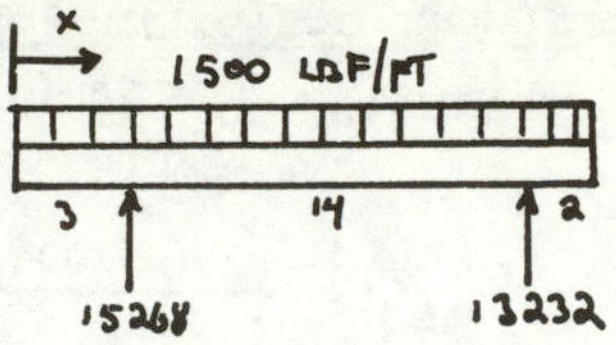

AT MID-SPAN,

$x = 10$

$V = 15268 - 10(1500)$

$= 268$ LBF

$M = 15268(7) - 1500(\frac{1}{2})(10)^2 = 31,876$ FT-LBF

$= 382,512$ IN-LBF

THE SHEAR STRESS IS

$$\tau_{max} = \frac{3V}{2A} = \frac{3(268)}{2(200)} = 2.01 \text{ PSI}$$

THE BENDING STRESS IS

$$\sigma_b = \frac{Mc}{I} = \frac{(382,512)(10)}{6666.7} = 574 \text{ psi}$$

15 FROM THE SHEAR AND MOMENT DIAGRAMS,

$V_{MAX} = 3916$ LBF

$M_{MAX} = 20058$ FT-LBF

$= 240696$ IN-LBF

$A = 5(10) = 50 \text{ IN}^2$

$I = \frac{5(10)^3}{12} = 416.7 \text{ IN}^4$

$$\tau_{MAX} = \frac{3V}{2A} = \frac{3(3916)}{2(50)} = 117.5 \text{ psi}$$

$$\sigma_b = \frac{Mc}{I} = \frac{240696(5)}{416.7} = 2888.1 \text{ psi}$$

16 FROM PAGE 11-21, (CASE 5)

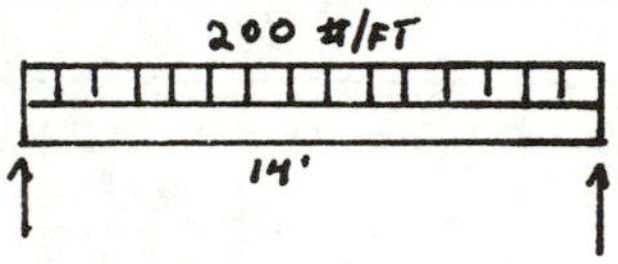

$$M_{MAX} = \frac{wL^2}{8} = \frac{200(14)^2}{8}$$

$= 4900$ FT-LBF

$= 58800$ IN-LBF

$I = \frac{3.625(7.625)^3}{12} = 133.9 \text{ IN}^4$

$$\sigma_b = \frac{Mc}{I} = \frac{58800(7.625/2)}{133.9} = 1674.2 \text{ psi}$$

17 $\sigma = \frac{Mc}{I} = \frac{M}{Z}$

$$Z = \frac{M_{MAX}}{\sigma_{MAX}} = \frac{(15000) \text{ FT-LBF } (12) \text{ IN/FT}}{2400 \text{ PSI}} = 7.5 \text{ IN}^3$$

18 TO FIND R_1 AND R_2, REPLACE THE LOAD WITH

$P = \frac{1}{2}(12)(w) = 6w$

P ACTS $(\frac{1}{3})(8+4) = 4$ FROM RIGHT END

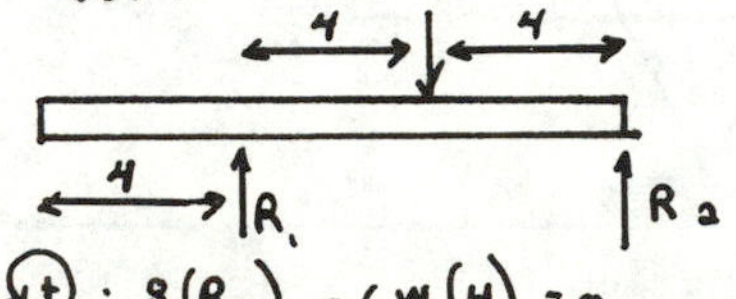

ΣM_{R_1} (+): $8(R_2) - 6w(4) = 0$

$R_2 = 3w$

AND $R_1 = 3w$

BETWEEN THE LEFT END AND R_1,

M_{max} OCCURS AT R_1

$M = \frac{1}{2}(4)(\frac{w}{3})(\frac{4}{3}) = 0.889w$ FT-LBF

BETWEEN R_1 AND R_2 (x MEASURED FROM THE LEFT END),

$M = \frac{1}{2}(x)(\frac{x}{12})w(\frac{x}{3}) - (x-4)3w$

$= \frac{x^3 w}{72} - 3xw + 12w$

$\frac{dM}{dx} = \frac{x^2 w}{24} - 3w = 0$

$\frac{x^2}{24} = 3$ OR $x = 8.49$

AT $x = 8.49$

$M = \frac{(8.49)^3 w}{72} - 3(8.49)w + 12w$

$= -4.97w$ FT-LBF $(> 0.889w)$

$\sigma = \frac{Mc}{I}$

$$22000 = \frac{(4.97)w(12)(5)}{78.5}$$

OR $w = 5791$ LBM/FT

19 THE DISTRIBUTED LOAD IS UNKNOWN.

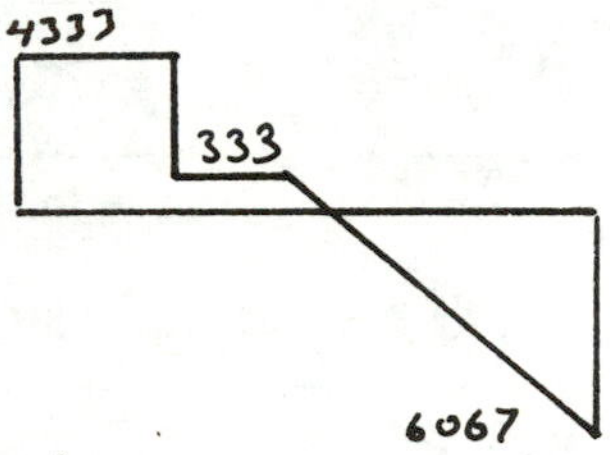

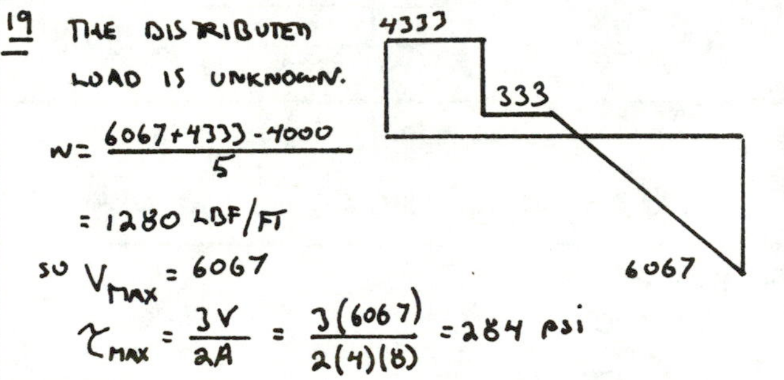

$$w = \frac{6067 + 4333 - 4000}{5}$$

$= 1280$ LBF/FT

SO $V_{MAX} = 6067$

$$\tau_{MAX} = \frac{3V}{2A} = \frac{3(6067)}{2(4)(8)} = 284 \text{ psi}$$

20 $A = 1 \text{ IN}^2$

$I = \frac{bh^3}{12} = .08333$

$M = Fe = 5000(1.5) = 7500$

$\sigma = \frac{F}{A} + \frac{Mc}{I}$

$= \frac{5000}{1} + \frac{7500(.5)}{.08333} = 50,000$ psi

21 $A = 2(\frac{3}{4}) = 1.5 \text{ IN}^2$

$I = \frac{bh^3}{12} = \frac{(.75)(2)^3}{12} = .5 \text{ IN}^2$

$M = 4000(1.25) = 5000$ IN-LBF

$\sigma = \frac{F}{A} + \frac{Mc}{I}$

$= \frac{4000}{1.5} + \frac{(5000)(1)}{.5}$

$= 12667$ psi

22 $A = 4(2\tfrac{1}{2}) = 10$

$I = \dfrac{(2.5)(4)^3}{12} = 13.33$

$M = 2000(2+2) = 8000$ IN-LBF

$\sigma = \dfrac{F}{A} + \dfrac{Mc}{I}$

$= \dfrac{2000}{10} + \dfrac{8000(2)}{13.33} = 1400$ psi

23 $\sigma_x = \dfrac{200{,}000}{(8)(4)} = -6250$ psi

{NEGATIVE BECAUSE IN COMPRESSION}

$\sigma_y = \dfrac{50{,}000}{(2)(4)} = -6250$ psi

FROM EQN 11.46, THE PRINCIPAL STRESSES ARE

$\tfrac{1}{2}(-6250 - 6250) \pm \tfrac{1}{2}\sqrt{(-6250-(-6250))^2 + (0)^2}$

$= -6250 \pm 0$

$= -6250, \ -6250$

24 $A = \frac{\pi}{4}(2)^2 = 3.142 \text{ IN}^2$

$J = \frac{\pi}{2}(r)^4 = \frac{\pi}{2}(1)^4 = 1.571$

$\sigma = \dfrac{F}{A} = \dfrac{-15{,}000}{3.142} = -4774$ PSI {COMPRESSIVE}

$\tau = \dfrac{Tc}{J} = \dfrac{(2500)(1)}{1.571} = 15913$

THE PRINCIPAL STRESSES ARE

$\sigma = \tfrac{1}{2}(-4774) \pm \tfrac{1}{2}\sqrt{(-4774)^2 + [(2)(15913)]^2}$

$= 13{,}704$ AND $-18{,}478$

25 $y = \dfrac{w}{48EI}(2x^4 - 5Lx^3 + 3L^2x^2)$

$\dfrac{dy}{dx} = \dfrac{w}{48EI}(8x^3 - 15Lx^2 + 6L^2x)$

$= \dfrac{w}{48EI}(x)(8x^2 - 15Lx + 6L^2)$

FACTORING THE QUADRATIC,

$x^2 - 1.875Lx + .75L^2 = 0$

$(x - .9375L)^2 = -.75L^2 + .879L^2$

$x - .9375L = \pm .359L$

$x = 1.3L, \ .578L$

$x = .578L$ IS THE ONLY NON-TRIVIAL, FEASIBLE SOLUTION

26 USE SUPERPOSITION,

$L = (17)\text{FT}(12)\text{IN/FT} = 204$

FROM THE DISTRIBUTED LOAD

FROM CASE (5) ON PAGE 11-21,

$y_{\text{mid-point}} = y_{max} = \dfrac{5\left(\frac{500}{12}\right)\frac{\text{LBF}}{\text{IN}}\left[(204)^4\right]}{(384)(3EE7)(200)}$

$= .1566''$

FROM THE CONCENTRATED LOAD

FROM CASE (11) PAGE 11-22,

$b = (5)(12) = 60$

$x = \left(\frac{17}{2}\right)12 = 102$

$y_{\text{midpoint}} = \dfrac{(2000)(60)(102)}{(6)(3EE7)(200)(204)}\left[(204)^2 - (60)^2 - (102)^2\right]$

$= .0460$

$y_{\text{total}} = .1566 + .0460 = .2026''$

27 $I_1 = \dfrac{4(12)^3}{12} = 576$

$I_2 = \dfrac{12(4)^3}{12} = 64$

$y_2 = \dfrac{576}{64}(.2) = 1.8''$

28 $A_1 = 2(10) = 20$

$\bar{x}_1 = 5$

$I_1 = \dfrac{2(10)^3}{12} = 166.67$

$A_2 = 2(5) = 10$

$\bar{x}_2 = 11$

$I_2 = \dfrac{5(2)^3}{12} = 3.33$

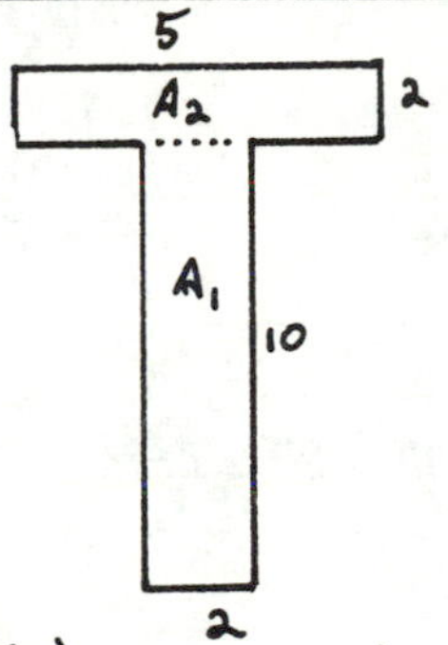

$\bar{x}_{\text{composite}} = \dfrac{(20)5 + (10)11}{20+10} = 7''$

$I_{\text{total}} = 166.67 + 20(7-5)^2 + 3.33 + 10(11-7)^2$

$= 410 \text{ IN}^4$

CONSIDER THE 4 FEET NEAREST THE WALL.

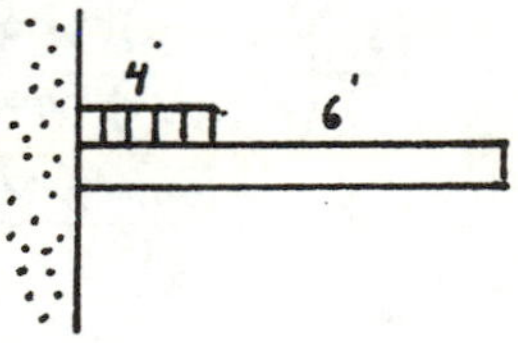

FROM PAGE 11-21 CASE (2)

$y = \dfrac{w}{24EI}(3L^4 - 4L^3x + x^4)$

$\dfrac{dy}{dx} = \dfrac{w}{24EI}(4x^3 - 4L^3)$

$y_{max} = \dfrac{wL^4}{8EI}$

AT $L = 48''$

$y = \dfrac{\left(\frac{8000}{12}\right)(48)^4}{(8)(3EE7)(410)} = .036$

AT $x = 0$ {MEASURED FROM THE RIGHT EDGE OF THE DISTRIBUTED LOAD}

$y' = \dfrac{\frac{8000}{12}}{(24)(3EE7)(410)}(.4(48)^3) = 1\text{ EE-3}$

THE DROP IN TIP OVER THE NEXT 6 FEET IS

$(6)(12)(1\text{ EE-3}) = 7.2\text{ EE-2}'$

so $y_{tip} = .036 + .072 = .108''$

29 $L = (10)(12) = 120$

$F_{SPRING} = .3(30000) = 9000$ LBF

USE SUPERPOSITION.

FROM PAGE 11-21 {CASE 4}

$$y = \frac{FL^3}{48EI} = \frac{(9000)(120)^3}{48(3EE7)(100)} = .108$$

FROM PAGE 11-21 {CASE 5}

$$y = \frac{5wL^4}{384EI} = \frac{5\left(\frac{w}{12}\right)(120)^4}{384(3EE7)(100)} = (7.5\,EE\text{-}5)\,w$$

SO $(7.5\,EE\text{-}5)w - .108 = .3$

$w = 5440$ LBF/FT

30 $J = \frac{\pi}{2}r^4 = \frac{\pi}{2}\left(\frac{2.5}{2}\right)^4 = 3.835$

$\tau = \frac{Tc}{J}$

SO $T = \frac{\tau J}{c} = \frac{(10,000)(3.835)}{\left(\frac{2.5}{2}\right)} = 30680$ IN-LBF

FROM EQUATION 11.76

$$\theta = \frac{TL}{GJ} = \frac{(30680)(2)(12)}{(11.5EE6)(3.835)} = .0167 \text{ RADIANS} = .96°$$

31 $J = \frac{\pi}{2}r^4 = \frac{\pi}{32}(d)^4$

AS LIMITED BY STRESS

$\tau = \frac{Tc}{J} = \frac{Td}{2J}$

AND $HP = \frac{T(RPM)}{63025}$

SO $J = \frac{Tc}{\tau} = \frac{(63025)(HP)(d)}{(RPM)(\tau)(2)}$

AND $d = \sqrt[3]{\frac{(32)(63025)(200)}{(\pi)(2)(1850)(10000)}} = 1.514''$

AS LIMITED BY DEFLECTION

$\theta = \frac{TL}{GJ}$ OR

$J = \frac{TL}{G\theta} = \frac{(63025)(HP)(L)}{(RPM)(G)(\theta)}$

$$d = \sqrt[4]{\frac{32(63025)(200)(1)(12)}{\pi(1850)(11.5EE6)\left(\frac{1}{360}\right)(2\pi)}} = 1.427''$$

SO, $d = 1.514''$

32 $J = \frac{\pi}{2}r^4 = \frac{\pi}{2}(1)^4 = 1.571$

$$\tau = \frac{Tc}{J} = \frac{(63025)(HP)(c)}{(RPM)J} = \frac{(63025)(200)(1)}{(875)(1.571)} = 9169.8 \text{ PSI}$$

33 $J = \frac{\pi}{2}r^4 = 1.571$

$$HP = \frac{T\,RPM}{63025} = \frac{\frac{\tau J}{c}(RPM)}{63025} = \frac{(12000)(1.571)(100)}{(1)(63025)}$$

$= 29.91$ HP

34 $I = \frac{4(4)^3}{12} = 21.33$

FROM TABLE 11.3, $C = 2$

SO, FROM EQN 11.58

$L' = CL = (2)(24)(12) = 576''$

FROM EQN. 11.50

$$F_e = \frac{\pi^2(1.5EE6)(21.33)}{(576)^2} = 951.8 \text{ LBF}$$

35 $I = \frac{\pi}{4}(r)^4 = \frac{\pi}{4}(.5)^4 = .0491$

$L' = (.5)(8)(12) = 48$ {C = .5 FROM TABLE 11.3}

$$F_e = \frac{(\pi)^2(3EE7)(.0491)}{(48)^2} = 6310 \text{ LBF}$$

$$\sigma = \frac{F}{A} = \frac{6310}{\frac{\pi}{4}(1)^2} = 8034$$

$\sigma = E\epsilon_{th} = E\alpha\Delta T$

SO $\Delta T = \frac{8034}{(3EE7)(6.5EE\text{-}6)} = 41.2°F$

36 $A = \frac{\pi}{4}\left[(3)^2 - (D_i)^2\right]$

AS LIMITED BY NORMAL STRESS

$\sigma = \frac{F}{A}$

SO $A = \frac{F}{\sigma}$

$$\frac{\pi}{4}\left[(3)^2 - (D_i)^2\right] = \frac{5000}{\left(\frac{22000}{2}\right)}$$

$D_i = 2.90$

$t = \frac{3.00 - 2.90}{2} = .05$

AS LIMITED BY BUCKLING

$I = \frac{\pi}{4}\left[(r_o)^4 - (r_i)^4\right] = \frac{\pi}{4}\left[(1.5)^4 - (r_i)^4\right]$

FROM TABLE 11.3, $C = 1$

SO $L = L' = (10)(12) = 120''$

$F_e = \frac{\pi^2 EI}{L}$

$$5000(2) = \frac{\pi^2(3EE7)\left(\frac{\pi}{4}\right)\left[(1.5)^4 - (r_i)^4\right]}{(120)^2}$$

$r_i = 1.452$

$t = 1.5 - 1.452 = .048''$

SO $t = .05$ {COMPRESSION GOVERNS}

37 a) FROM EQN 11.62

$$\sigma_h = \frac{pr}{t} = \frac{(1900)(3/2)}{(1/8)} = 38000 \text{ PSI}$$

b) FROM EQN 11.63

$\sigma_\ell = \frac{pr}{2t} = 19000$ PSI

{MORE}

PROBLEM 37, CONTINUED

c) THE PRINCIPAL STRESSES ARE

$$\frac{1}{2}(38000+19000) \pm \frac{1}{2}\sqrt{(38000-19000)^2+0}$$

$$= 28500 \pm 9500$$

$$= 19000,\ 38000$$

38 SINCE NO THICK-WALL CYLINDERS HAVE APPEARED ON THE E-I-T EXAM, ASSUME THIN-WALL EQUATIONS CAN BE USED. USE EQN 11.62

$$p = \frac{F}{A} = \frac{100{,}000}{\frac{\pi}{4}(10)^2} = 1273 \text{ PSI}$$

$$\sigma_h = \frac{(1273)(10/2)}{1} = 6365 \text{ psi}$$

39 $F = KX$

$$x = \frac{18}{12} = 1.5'$$

40

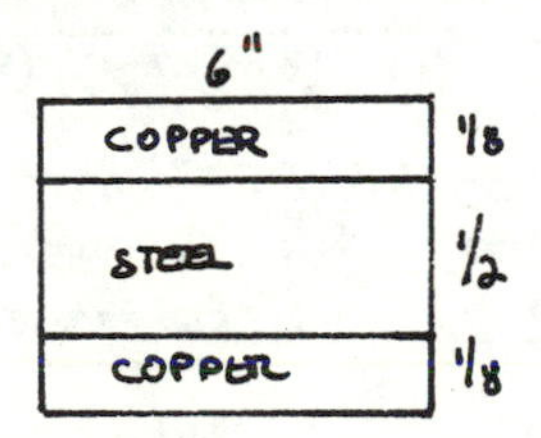

USE THE PROCEDURE ON PAGE 11-7

$$h = \frac{E_{STEEL}}{E_{COPPER}} = \frac{3\ EE7}{1.75\ EE7} = 1.714$$

$$(6)(1.714) = 10.284$$

THE TRANSFORMED BAR CROSS SECTION LOOKS LIKE

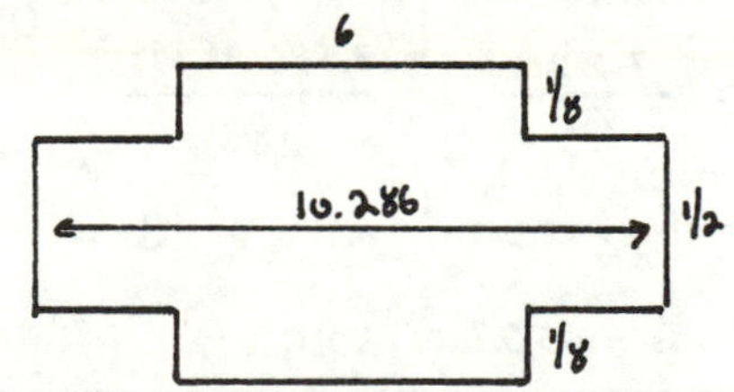

THE TOTAL (ALL COPPER) AREA IN COMPRESSION IS

$$2\left(\tfrac{1}{8}\right)(6) + (6)\left(\tfrac{1}{2}\right)(1.714) = 6.642 \text{ IN}^2$$

$$\sigma_{COPPER} = \frac{F}{A} = \frac{100{,}000}{6.642} = 15{,}056 \text{ PSI}$$

$$\sigma_{STEEL} = (1.714)(\sigma_{COPPER}) = 25{,}805$$

$$\epsilon = \frac{\sigma}{E} = \frac{15056}{1.75EE7} = .00086$$

41 USE THE TRANSFORMATION METHOD

$$n = \frac{E_{STEEL}}{E_{COPPER}} = \frac{3\ EE7}{1.75\ EE7} = 1.714$$

$$A_{COPPER} = \frac{\pi}{4}\left[(10)^2 - (5)^2\right] = 58.9$$

$$A_{STEEL} = \frac{\pi}{4}(5)^2 = 19.63$$

TRANSFORMED AREA $= 58.9 + 1.714(19.63)$
$= 92.55 \text{ IN}^2$

$$\sigma_{COPPER} = \frac{F}{A} = \frac{100{,}000}{92.55} = 1080.5 \text{ PSI}$$

$$\sigma_{STEEL} = 1.714\,\sigma_{COPPER} = 1852 \text{ psi}$$

42 USE THE TRANSFORMATION METHOD

$$h = \frac{E_{STEEL}}{E_{COPPER}} = \frac{3\ EE7}{1.75\ EE7} = 1.714$$

TOTAL TRANSFORMED AREA IS

$$.6 + (2)(1.714)(.2) = 1.2856$$

$$\sigma_{COPPER} = \frac{F}{A} = \frac{2(10{,}000)}{1.2856} = 15557 \text{ psi}$$

$$F_{COPPER} = \sigma A = (15557)(.6) = 9334 \text{ LBF}$$

$$\sigma_{STEEL} = 1.714\,\sigma_{COPPER} = 1.714(15557) = 26665 \text{ PSI}$$

$$F_{STEEL} = \sigma A = (26665)(.2) = 5333 \text{ LBF}$$

43 THE FORCE AT C IS

$$20000\left(\tfrac{1}{2}\right) = 10{,}000$$

USE THE TRANSFORMATION METHOD

$$h = \frac{E_{STEEL}}{E_{COPPER}} = \frac{3\ EE7}{1.75\ EE7} = 1.714$$

THE TRANSFORMED SUPPORTING AREA IS

$$1 + 1.714(4) = 7.856$$

$$\delta = \frac{FL}{AE} = \frac{(10{,}000)(4)(12)\ \text{in/FT}}{(7.856)(1.75\ EE7)} = 3.49\ EE\text{-}3''$$

E-I-T HOMEWORK SOLUTIONS: D. C. ELECTRICITY

1 FROM PAGE 1-36

$$E = \frac{(5)\ \text{HP}\ (.7457)\ \text{KW/HP}\ (1)\ \text{HR/DAY}\ (15)\ \text{DAYS}}{.80}$$

$$+ \frac{(500)\ \text{WATT}\ (4)\ \text{HR/DAY}\ (30)\ \text{DAYS}}{(1000)\ \text{W/KW}} = 129.9\ \text{KW-HRS}$$

2 $(2500)\ \text{gpm}\ (.002228)\ \frac{\text{FT}^3}{\text{SEC-gpm}} = 5.57\ \text{FT}^3/\text{sec}$

$\dot{m} = Q\rho = (5.57)\ \text{FT}^3/\text{sec}\ (62.4)\ \text{LBM/FT}^3$

$= 347.57\ \text{LBM/sec}$

$P = \dot{m}\Delta h = (347.57)\ \frac{\text{LBM}}{\text{SEC}}\ (120)\ \text{FT} = 41708.4\ \frac{\text{FT-LBF}}{\text{sec}}$

$\text{HP} = \frac{P}{550\ \eta} = \frac{41708.4}{(550)\ (.70)} = 108.3$

$\text{KW} = (108.3)\ \text{HP}\ (.7457)\ \frac{\text{KW}}{\text{HP}} = 80.8\ \text{KW}$

3 $v = \frac{(7)\ \text{MPH}\ (5280)\ \text{FT/MI}}{(3600)\ \text{SEC/HR}} = 10.267\ \text{FT/sec}$

$P = Fv = (200)\ \text{LBF}\ (10.267)\ \frac{\text{FT}}{\text{SEC}} = 2053.4\ \frac{\text{FT-LBF}}{\text{SEC}}$

FROM PAGE 7-1

$(2053.4)\ \frac{\text{FT-LBF}}{\text{SEC}}\ (1.356\ \text{EE-3})\ \frac{\text{KW-sec}}{\text{FT-LBF}}\ (1000)\ \text{W/KW}$

$= 2784.4\ \text{W}$

FROM EQN 13.61

$P = IV$, SO $I = \frac{P}{V} = \frac{2784.4\ \text{W}}{(.75)(110)\ \text{V}} = 33.75\ \text{A}$

4 $P = IV$ OR

$I = \frac{P}{V} = \frac{(800)\ \text{W}}{(110)\ \text{V}} = (7.273)\ \text{AMPS}$

FROM EQN 13.53

$(7.273)\ \text{AMP}\ (1)\ \frac{\text{coul}}{\text{SEC-AMP}}\ (6.24\ \text{EE18})\ \frac{\text{ELECTRONS}}{\text{coul}}$

$= 4.538\ \text{EE19}\ \frac{\text{ELECTRONS}}{\text{SEC}}$

5 a) FROM EQN 13.8

$$F = \frac{(-1\ \text{EE-6})\ \text{coul}\ (-2\ \text{EE-6})\ \text{coul}}{(4\pi)\ (8.85\ \text{EE-12})\ \frac{\text{coul}^2}{\text{N-M}^2}\ (1)^2\ \text{M}^2} = 1.8\ \text{EE-2}\ \text{N}$$

b) FORCE IS ALONG THE Y-AXIS

c) FROM EQN 13.10

$$E_p = \frac{-(-1\ \text{EE-6})\ (-2\ \text{EE-6})}{(4\pi)\ (8.85\ \text{EE-12})\ (1)} = -1.8\ \text{EE-2}\ \text{J}$$

FROM EQN 13.11

$$V_2 = \frac{E_p}{q_2} = \frac{-(1.8\ \text{EE-2})\ \text{J}}{(-2\ \text{EE-6})\ \text{C}} = 9000\ \text{V}$$

6 PT 1 $= (2,0,0)$

PT 2 $= (0,2,0)$

$d_{1-2} = \sqrt{(2-0)^2 + (0-2)^2 + (0-0)^2} = 2.828\ \text{M}$

a) FROM EQN 13.8

$$F = \frac{(4\ \text{EE-6})\ (-5\ \text{EE-6})}{(4\pi)\ (8.85\ \text{EE-12})\ (2.828)^2} = -.02249\ \text{N}$$

b) PT 3 $= (3,5,2)$

$d_{1-3} = \sqrt{(3-2)^2 + (5-0)^2 + (2-0)^2} = 5.477\ \text{M}$

$d_{2-3} = \sqrt{(3-0)^2 + (5-2)^2 + (2-0)^2} = 4.690\ \text{M}$

THE WORK TO BRING q_3 FROM ∞ TO WITHIN 5.477 M OF q_1 IS {EQN 13.10}

$$E_{p,1,3} = \frac{-(4\ \text{EE-6})\ q_3}{(4\pi)\ (8.85\ \text{EE-12})\ (5.477)}$$

$= -(6.567\ \text{EE3})\ q_3$

THE POTENTIAL AT PT 3 DUE TO q_1 IS

$$V_1 = \frac{E_{p,1-3}}{q_3} = \frac{-(6.567\ \text{EE3})\ q_3}{q_3} = -6.567\ \text{EE3}$$

THE WORK TO BRING q_3 FROM ∞ TO WITHIN 4.690 M OF q_2 IS

$$E_{p,2-3} = \frac{-(-5\ \text{EE-6})\ q_3}{(4\pi)\ (8.85\ \text{EE-12})\ (4.690)} = (9.586\ \text{EE3})\ q_3$$

THE POTENTIAL AT POINT 3 DUE TO q_2 IS

$$V_2 = \frac{E_{p,2-3}}{q_3} = \frac{(9.586\ \text{EE3})\ q_3}{q_3} = 9.586\ \text{EE3}$$

THE TOTAL POTENTIAL AT PT 3 IS

$9586 - 6567 = 3019\ \text{V}$

7 FROM EQN 13.5

$$E = \frac{(.5\ \text{EE-6})}{(4\pi)(8.85\ \text{EE-12})\ (2)^2} = 1124\ \frac{\text{VOLTS}}{\text{METER}}$$

$= 1124\ \text{N/C}$

8 $R_{C-A} = (0-0)i + (3-0)j = 3j$

$|R_{C-A}| = \sqrt{(0-0)^2 + (3-0)^2} = 3$

$a_{r,C-A} = \frac{3j}{3} = j$

$$E_{C-A} = \frac{2\ \text{EE-6}}{4\pi\ (8.85\ \text{EE-12})(3)^2}\ (j) = 1998.2\ j$$

$R_{C-B} = (0-4)i + (3-0)j = -4i + 3j$

$|R_{C-B}| = \sqrt{(-4)^2 + (3)^2} = 5$

C, A, B, 2μC, -5μC

{MORE}

PROBLEM 8 CONTINUED

$$a_{r,C\text{-}B} = \frac{-4i+3j}{5} = -.8i + .6j$$

$$E_{C\text{-}B} = \frac{(5\text{EE}6)}{(4\pi)(8.85\text{EE-}12)(5)^2}(-.8i+.6j)$$

$$= 1438.7i - 1079j$$

$$E_{total} = E_{C\text{-}A} + E_{C\text{-}B} = (1438.7)i + (1998.2 - 1079)j$$

$$= (1438.7)i + (919.2)j$$

$$|E| = \sqrt{(1438.7)^2 + (919.2)^2} = 1707.3 \text{ N/C}$$

9 THE COORDINATES OF POINT C ARE (4,-3)

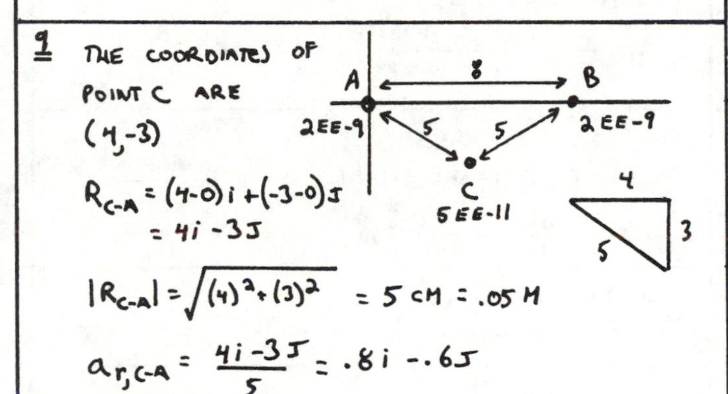

$$R_{C\text{-}A} = (4-0)i + (-3-0)j = 4i - 3j$$

$$|R_{C\text{-}A}| = \sqrt{(4)^2 + (3)^2} = 5 \text{ CM} = .05 \text{ M}$$

$$a_{r,C\text{-}A} = \frac{4i-3j}{5} = .8i - .6j$$

FROM EQN 13.8

$$F_{C\text{-}A} = \frac{(2\text{EE-}9)(5\text{EE-}11)}{(4\pi)(8.85\text{EE-}12)(.05)^2}(.8i - .6j)$$

$$= (2.877\text{EE-}7)i - (2.158\text{EE-}7)j$$

BY SYMMETRY

$$F_{C\text{-}B} = (-2.877\text{EE-}7)i - (2.158\text{EE-}7)j$$

$$F_C = F_{C\text{-}A} + F_{C\text{-}B} = (2.877\text{EE-}7 - 2.877\text{EE-}7)i$$

$$- (2.158\text{EE-}7 + 2.158\text{EE-}7)j$$

$$= -(4.316\text{EE-}7)j \text{ N (NEGATIVE = DOWN)}$$

10 a) FROM EQN 13.14

$$E = \frac{V}{r} = \frac{100 \text{ V}}{.005} = 20{,}000 \text{ V/M} = 20{,}000 \text{ N/C}$$

b) FROM EQN 13.13

$$E_{STORED} = \left(\tfrac{1}{2}\right)(8.85\text{EE-}12)\frac{\text{coul}^2}{\text{N-M}^2}(20{,}000)^2(\text{N/C})^2$$

$$= 1.77\text{EE-}3 \text{ N/M}^2$$

BUT A NEWTON IS A $\frac{\text{JOULE}}{\text{METER}}$

SO $E = 1.77\text{EE-}3$ joule/m^3

11 FROM EQN 13.13

$$V = Er = (\text{EE}6)\frac{\text{VOLT}}{\text{METER}}(.005)\text{M} = 5000 \text{ V}$$

12 a)

$$E = \frac{V}{r} = \frac{1200}{.5} = 2400 \text{ V/M}$$

$$F = Eq = (2400)\frac{\text{V}}{\text{M}}(1.602\text{EE-}19)\text{C}$$

$$= 3.845\text{EE-}16 \text{ N}$$

b) WORK TO MOVE THE ELECTRON FROM THE POSITIVE TO THE NEGATIVE PLATE IS GIVEN BY EQN 13.16

$$W = Fd = (3.845\text{EE-}16)(.5) = 1.923\text{EE-}16 \text{ J}$$

THIS IS ALSO THE POTENTIAL ENERGY OF THE SYSTEM, FROM EQN 13.11

$$V_1 = \frac{E_P}{q} = \frac{1.923\text{EE-}16 \text{ J}}{1.602\text{EE-}19 \text{ C}} = 1200 \text{ V} \checkmark$$

13 FROM EQN 13.20

$$v = (5.93\text{EE}5)\sqrt{18} = 2.516\text{EE}6 \text{ M/S}$$

14

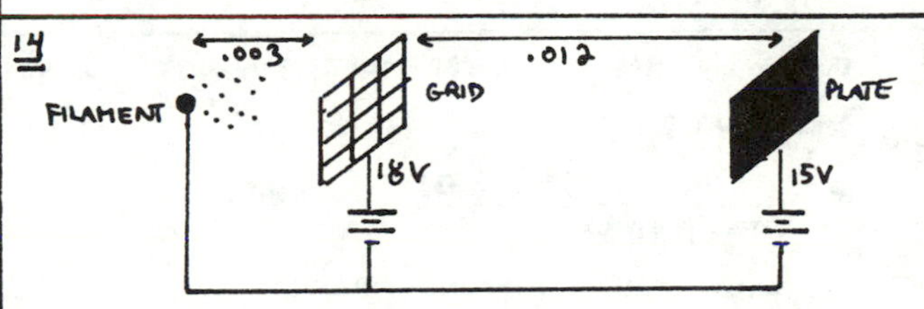

THE ELECTRICAL FIELD AT THE FILAMENT DUE TO THE 2 BATTERIES IS

$$E = \frac{18}{.003} + \frac{15}{(.003+.012)} = 7000 \text{ V/M}$$

AN ELECTRON STARTING WITH $V_0 = 0$ AT THE FILAMENT WILL ACHIEVE A VELOCITY AT THE GRID OF {EQN 13.19}

$$v = \sqrt{2Eqd/m} = \sqrt{\frac{(2)(7000)(1.602\text{EE-}19)(.003)}{(9.11\text{EE-}31)\text{ KG}}}$$

$$= 2.718\text{EE}6 \text{ M/S}$$

AFTER PASSING THROUGH THE GRID, THE FIELD IS

$$\frac{15}{.003+.012} + \frac{(15-18)}{.012} = 750 \text{ V/M}$$

ALSO FROM EQN 13.19

$$v = \sqrt{\frac{(2)(750)(1.602\text{EE-}19)(.012)}{9.11\text{EE-}31} + (2.718\text{EE}6)^2}$$

$$= 3.249\text{EE}6 \text{ M/sec}$$

15 REFER TO FIGURE 13.3

a) FROM EQN 13.22

$$E_d = \frac{V_d}{L_3} = 20{,}000 \text{ V/M}$$

SO

$$\theta = \text{ARCTAN}\left(\frac{(20000)(1.602\text{EE-}19)(.04)}{(9.11\text{EE-}31)(2\text{EE}7)^2}\right) = 19.38°$$

b) FROM EQN 13.21,

$$y = (.12 + .02)\text{TAN}(19.38) = .0492 \text{ M}$$

16 FROM EQN 13.26

$$B = \frac{\phi}{A} = \frac{(34\ EE\text{-}3)\ wb}{.02} = 1.7\ \frac{wb}{m^2} = 1.7\ \text{TESLA}$$

17 $B = 1.2\ wb/m^2$

THE POLE STRENGTH IS (EQNS 13.25 + 13.26)

$$M = \phi = BA = (1.2)\frac{wb}{m^2}\left(\frac{\pi}{4}\right)(.01)^2 = 9.42\ EE\text{-}5\ wb$$

THE MAGNETIC MOMENT IS

$$ML = (9.42\ EE\text{-}5)(.1) = 9.42\ EE\text{-}6\ wb\text{-}M$$

18 $M = 4$

$F = 3N$

FROM EQN 13.29

$$H = \frac{F}{M} = \frac{3N}{4\ wb} = .75\ N/wb$$

19 FROM EQN 13.29

$$F = \frac{(8)(8)}{(4\pi)(4\pi\ EE\text{-}7)(.1)^2} = 4.05\ EE8\ N$$

20 BETWEEN THE 2 LEFT-MOST MAGNETS, USING EQN 13.29

$$F = \frac{(10\ EE\text{-}3)(6\ EE\text{-}3)}{(4\pi)(4\pi\ EE\text{-}7)(.2)^2} = 94.99\ N$$

BETWEEN THE 2 END MAGNETS

$$F = \frac{(10\ EE\text{-}3)(-20\ EE\text{-}3)}{(4\pi)(4\pi\ EE\text{-}7)(.6)^2} = -35.18\ N$$

$F_t = 94.99 - 35.18 = 59.81\ N$ REPULSIVE TO LEFT

21 PT A: (0,0)

PT B: (.05,0)

PT C: (.032, .024)

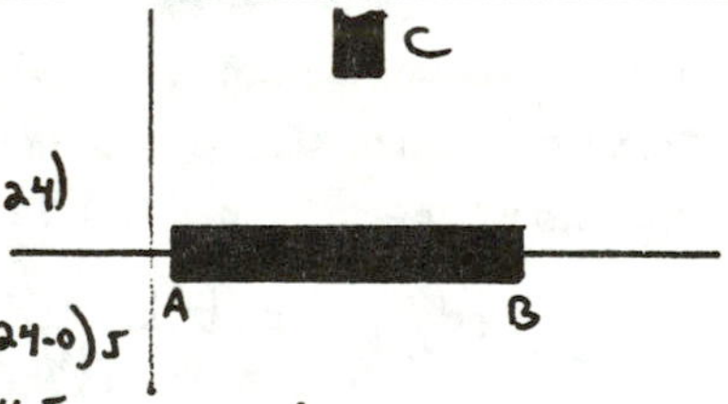

$R_{C\text{-}A} = (.032-0)i + (.024-0)J$

$= .032i + .024J$

$|R_{C\text{-}A}| = \sqrt{(.032)^2 + (.024)^2}$

$= .04$

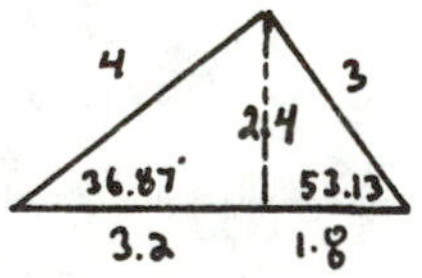

$$a_{r,C\text{-}A} = \frac{.032i + .024J}{.04}$$

$= .8i + .6J$

$$F_{C\text{-}A} = \frac{(15)(15)}{(4\pi)(4\pi\ EE\text{-}7)(.04)^2}(.8i + .6J)$$

$= (7.12\ EE9)i + (5.34\ EE9)J$

$R_{C\text{-}B} = (.032-.05)i + (.024-0)J = (-.018)i + (.024)J$

$|R_{C\text{-}B}| = .03$

$$a_{r,C\text{-}B} = \frac{-.018i + .024J}{.03} = -.6i + .8J$$

$$F_{C\text{-}B} = \frac{(15)(-.15)}{(4\pi)(4\pi\ EE\text{-}7)(.03)^2}(-.6i + .8J)$$

$= (9.5\ EE9)i - (1.27\ EE10)J$

$F_C = F_{CA} + F_{CB}$

$= (7.12\ EE9 + 9.5\ EE9)i + (5.34\ EE9 - 1.27\ EE10)J$

$= (1.66\ EE10)i - (7.36\ EE9)J$

$F_C = \sqrt{(1.66\ EE10)^2 + (7.36\ EE9)^2}$

$= 1.82\ EE10\ N$

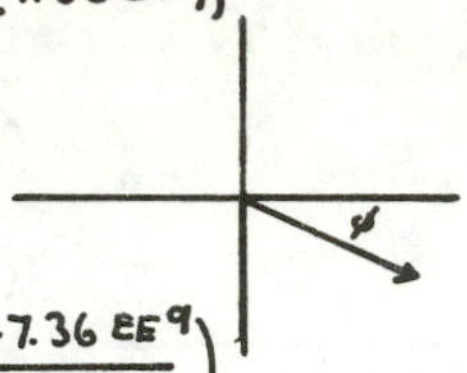

$$\phi = -23.9° = \text{ARCTAN}\left(\frac{-7.36\ EE9}{1.66\ EE10}\right)$$

FROM EQN 13.29

$$H = \frac{F}{M} = \frac{(1.82\ EE10)\ N}{(15)\ \text{UNIT-POLE}} = 1.21\ EE9\ N/wb$$

22 FROM EQN 13.45

$$F = IBL = (6)\ \text{AMP}\ (1.1)\ wb\ (.3)\ m = 1.98\ \frac{\text{AMP-wb}}{M}$$

BUT $\frac{\text{AMP}}{M} = \frac{N}{wb}$, SO 1.98 N

23 FROM EQN 13.33

$$\phi_2 = 30\ mwb\left(\frac{4}{5}\right) = 24\ mwb$$

24 FROM EQN 13.39

$$V = N\frac{d\phi}{dt} = 100\left(\frac{(40)\ wb/min}{(60)\ sec/min}\right) = 66.7\ V$$

25 FROM EQN 13.42

$$V = BLv = (4)\frac{wb}{m^2}(.10)\ m\ (1)\ m/sec = .4\ V$$

26 FROM EQN 13.50

$MMF = (.9)(400) = 360$ AMP TURNS

27 USE THE CENTERLINE DISTANCES TO CALCULATE RELUCTANCE.

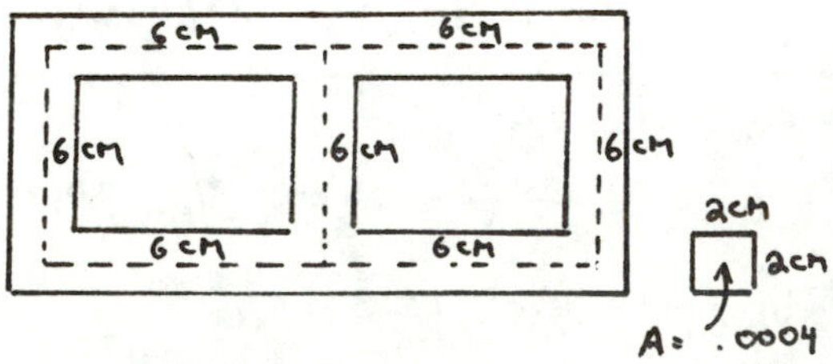

$\mu_r = 8400$ (GIVEN)

FROM EQN 13.51

$$\mathcal{R}_{LEG} = \frac{(.06)}{(8400)(4\pi\ EE\text{-}7)(.0004)} = 1.42\ EE4$$

THIS MAGNETIC CIRCUIT CAN BE CONSIDERED AS AN ELECTRICAL CIRCUIT:

(MORE)

PROBLEM 27 CONTINUED

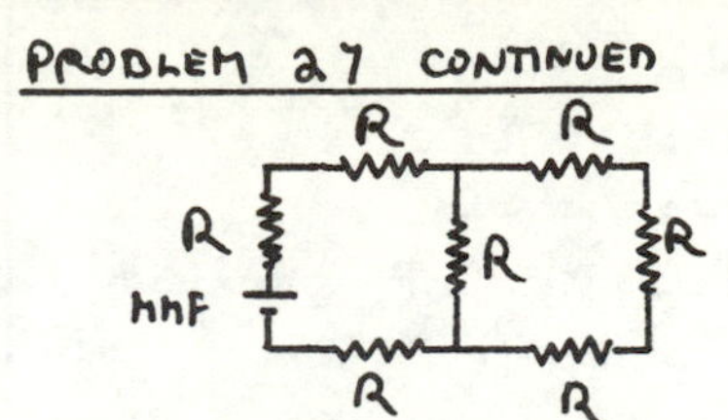

SIMPLIFYING,

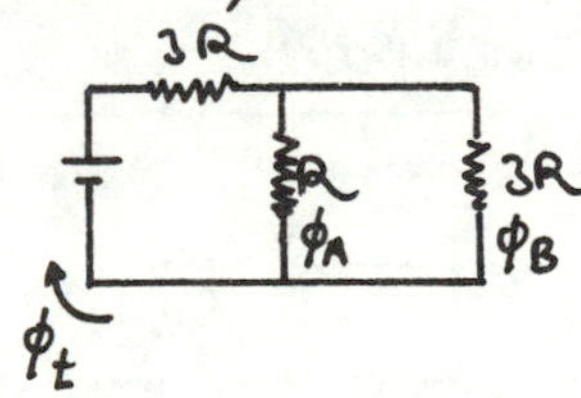

BECAUSE THE RIGHT BRANCH RELUCTANCE IS 3 TIMES THE CENTER BRANCH,

$$\phi_B = \frac{1}{3}\phi_A$$

AND $\phi_t = \phi_A + \phi_B = \frac{4}{3}(\phi_A) = \frac{4}{3}(3.6\ \text{EE-4})$

$= 4.8\ \text{EE-4 wb}$

FURTHER SIMPLIFICATION YIELDS

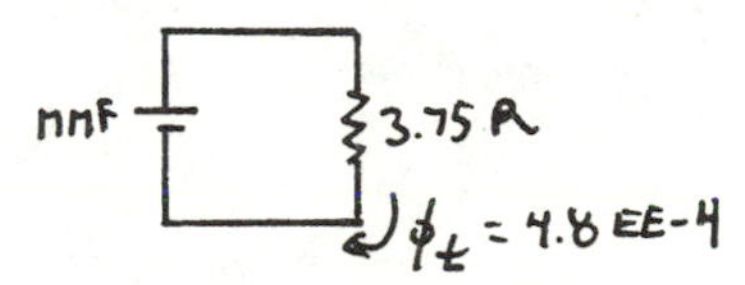

FROM EQN 13.52

$\text{MMF} = \phi R$

$(300)I = (4.8\ \text{EE-4})(3.75)(1.42\ \text{EE4})$

$I = .0852\ \text{A}$

28 USE CENTERLINE DISTANCES TO FIND RELUCTANCE

(40, 19.5, 40, 19.5, 40, 40)

$$\Sigma R = \Sigma \frac{L}{\mu A}$$

$$= \frac{.4 + .4 + .4 + .195 + .195}{(800)(4\pi\ \text{EE-7})\ (.10)^2} + \frac{.01}{(4\pi\ \text{EE-7})(.10)^2} = 9.54\ \text{EE5}$$

FROM TABLE 13.8

600 KILOLINES $= (600\ \text{EE3})(\text{EE-8}) = 6\ \text{EE-3 wb}$

FROM EQN 13.52

$5(N) = (6\ \text{EE-3})(9.54\ \text{EE5})$

$N = 1145$ TURNS

29 FROM PAGE 13-9 $\rho_0 = 10.37$ OHMS-CM/FT AND $\alpha = .00393\ 1/°\text{C}$

a) FROM EQN 13.56

$$\text{CM} = \left(\frac{.064}{.001}\right)^2 = 4096$$

$$R_0 = \frac{\rho_0 L}{A} = \frac{(10.37)(500)}{4096} = 1.266\ \Omega$$

b) $R = 1.266\left[1 + (.00393)(80-20)\right] = 1.565\ \Omega$

30 $$\alpha = \frac{R_2 - R_1}{R_1(T_2 - T_1)} = \frac{400 - 30}{30(200-100)} = .123\ /°\text{C}$$

THIS USES 100 °C AS THE REFERENCE TEMPERATURE.

31 THE INPUT POWER IS 2000 W. THE CURRENT IS

$$I = \frac{P}{V} = \frac{2000}{240} = 8.333\ \text{A}$$

THE ACCEPTABLE LINE LOSS IS

$(.05)(2000) = 100\ \text{W}$

THE ACCEPTABLE RESISTANCE IS

$$R = P/I^2 = \frac{100}{(8.333)^2} = 1.44\ \Omega$$

BUT $R = \frac{\rho L}{A}$ ALSO.

$$1.44 = \frac{(.6788\ \text{EE-6})\frac{\text{OHM-IN}^2}{\text{IN}}(2)\frac{\text{KM}}{\text{ROUND TRIP}}(3281)\frac{\text{FT}}{\text{KM}}(12)\frac{\text{IN}}{\text{FT}}}{\left(\frac{\pi}{4}\right)D^2}$$

SO $D = .217$ IN

32 $$\frac{\Delta R}{\Delta T} = \frac{(12-11)}{(31-5)} = .03846\ \Omega/°\text{C}$$

TO DROP FROM 11 Ω TO 0 WOULD REQUIRE A TEMPERATURE OF

$$5°\text{C} - \frac{11}{.03846} = -281\ °\text{C}$$

33 FROM EQN 13.61

$P = I^2R = (.5)^2 5 = 1.25\ \text{W}$

34 FROM EQN 13.62

$$\text{dB} = 10 \log\left(\frac{100\ \text{EE-3}}{400}\right) = -36.02\ \text{dB LOSS}$$

35 THE EQUIVALENT RESISTANCE IS

$$6 + 5 + \frac{(5)(8)}{5+8} = 14.08\ \Omega$$

THE TOTAL POWER DISSIPATED IS

$$P = \frac{V^2}{R} = \frac{(24+6)^2}{14.08} = 63.9\ \text{W}$$

$$P_{24} = \frac{24}{24+6}(63.9) = 51.12\ \text{W}$$

$$P_6 = \frac{6}{24+6}(63.9) = 12.78\ \text{W}$$

36 $V = (100 - 66) = 34\ \text{V}$

$R = 1 + 5 + 2 = 8\ \Omega$

$I = 34/8 = 4.25$

(1 Ω, 5 Ω, 100V, I, 66V, 2 Ω)

a) $P = \frac{V^2}{R} = \frac{(34)^2}{8} = 144.5\ \text{W}$

b) $P = (4.25)(66) = 280.5\ \text{W}$

$$E = \frac{(280.5)\ \text{W}}{(1000)\ \text{W/KW}}(1)\ \text{HR} = .2805\ \text{KW-HRS}$$

37 $100 + 25 + 9 = 134\ \text{dB}$

38

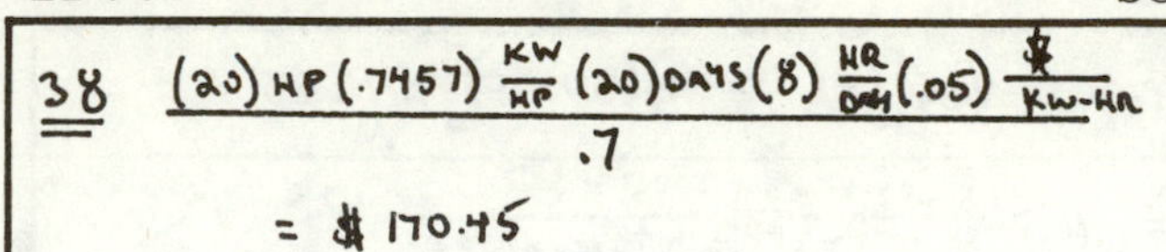

$$\frac{(20)\,HP\,(.7457)\,\frac{KW}{HP}\,(20)\,DAYS\,(8)\,\frac{HR}{DAY}\,(.05)\,\frac{\$}{KW\text{-}HR}}{.7}$$

$= \$\,170.45$

39

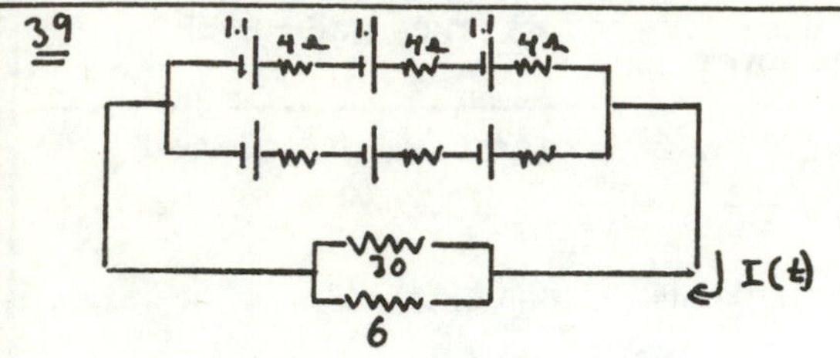

$V_{EQUIV} = 3(1.1) = 3.3$

$R_{LOAD} = \dfrac{(30)(6)}{36} = 5$

$R_{BATTERY} = \dfrac{(12)(12)}{12+12} = 6\,\Omega$

$R_t = 5 + 6 = 11$

$I = \dfrac{V}{R} = \dfrac{3.3}{11} = .3$

40

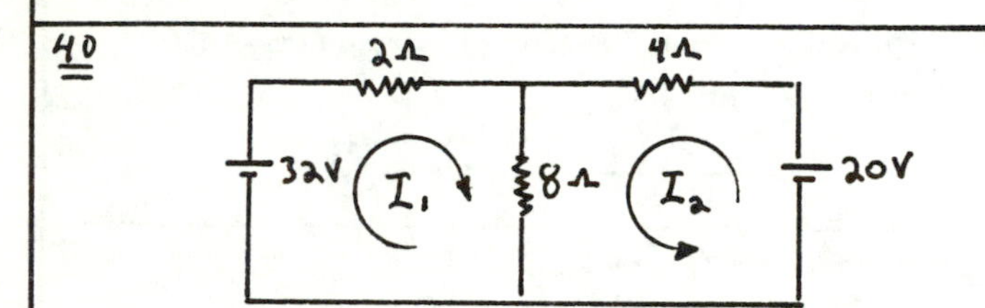

USE THE PROCEDURE ON PAGE 13-11

$32 = 2I_1 + 8(I_1 + I_2)$

$20 = 4I_2 + 8(I_1 + I_2)$

OR

$32 = 10I_1 + 8I_2$

$20 = 8I_1 + 12I_2$

THIS HAS SOLUTION OF

$I_1^* = 4 \qquad I_2^* = -1$

SO $I_{CENTER} = 4 - 1 = 3$ AMPS

41 CONSIDER THE FOLLOWING LOOP CURRENTS

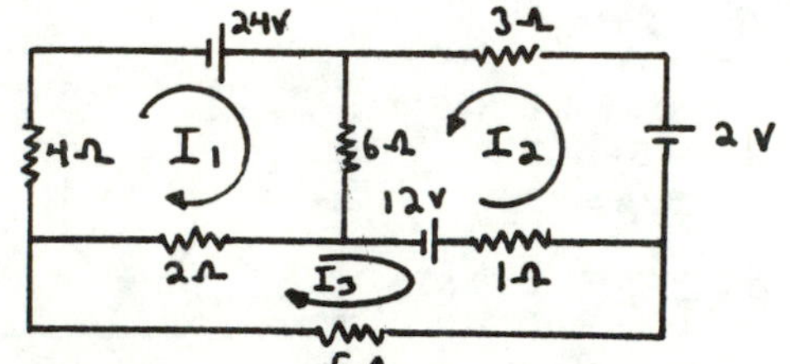

THE LOOP EQUATIONS ARE

$24 = 6(I_1 + I_2) + 2(I_1 - I_3) + 4I_1$

$12 + 2 = 3(I_2) + 6(I_1 + I_2) + (I_2 + I_3)$

$12 = 5I_3 + 2(I_3 - I_1) + I_3 + I_2$

THESE SIMPLIFY TO

$24 = 12I_1 + 6I_2 - 2I_3$

$14 = 6I_1 + 10I_2 + I_3$

$12 = -2I_1 + I_2 + 8I_3$

THE SOLUTION TO THIS SET OF EQUATIONS IS

$I_1^* = 2.53$

$I_2^* = -.336$

$I_3^* = 2.17$

$I_{6\Omega} = 2.530 - .336 = 2.194\,A$

42 CONSIDER THE FOLLOWING LOOP CURRENTS

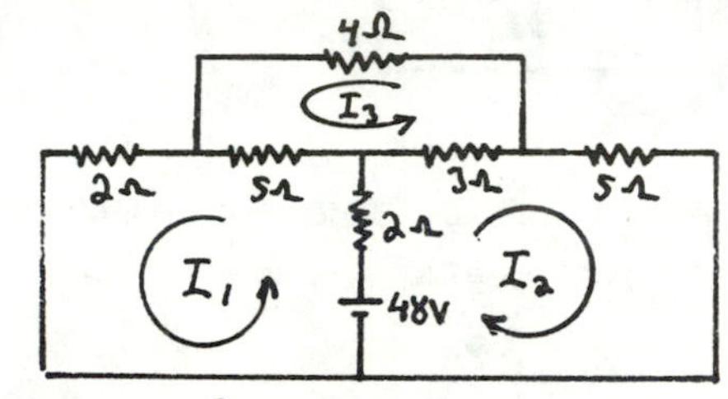

THE LOOP EQN'S ARE

$48 = 2I_1 + 5(I_1 - I_3) + 2(I_1 + I_2)$

$48 = 5I_2 + 3(I_2 + I_3) + 2(I_1 + I_2)$

$0 = 5(I_3 - I_1) + 3(I_3 + I_2) + 4I_3$

$I_1^* = 5.39$

$I_2^* = 3.41$

$I_3^* = I_{4\Omega} = 1.423$

43 FROM PAGE 13-12

$R = \dfrac{V}{I} = \dfrac{50}{8} = 6.25$

THEVENIN

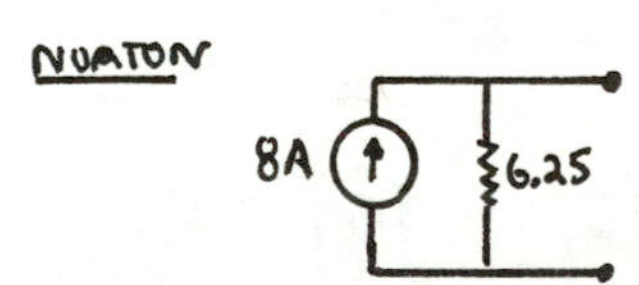

NORTON

8A

6.25

44 ASSUME LINEARITY WITHIN THE WORKING REGION

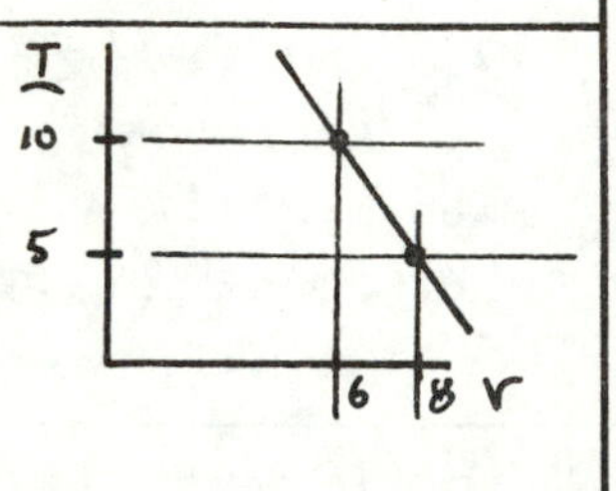

$SLOPE = \dfrac{\Delta I}{\Delta V} = \dfrac{10-5}{6-8}$

$= -2.5$

THE INTERCEPT IS

$10 = -2.5(6) + I_o \qquad I_o = 25$

SO $I = -2.5V + 25$

SINCE $R = \dfrac{V}{I} = \dfrac{V}{-2.5V + 25}$,

(MORE)

PROBLEM 44 CONTINUED

THEVENIN

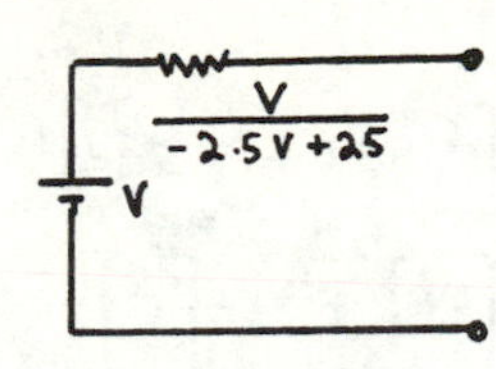

NORTON

$V = 10 - .4I$

$R = \frac{V}{I}$

$= \frac{10}{I} - .4$

I

$\frac{10}{I} - .4$

---

45 FROM EQNS 13.76-13.79

$R = 30 + 10 + 20 = 60$

$R_1 = \frac{(30)(20)}{60} = 10\ \Omega$

$R_2 = \frac{(30)(10)}{60} = 5\ \Omega$

$R_3 = \frac{(10)(20)}{60} = 3.33\ \Omega$

---

46 FROM EQN 13.84

$R_s = \frac{(.001)(500)}{.1 - .001} = 5.05\ \Omega$

---

47 FROM EQN 13.85

$R_M = \frac{100 - (.001)(500)}{.001} = 99{,}500\ \Omega$

---

48 a) $I_c^* = \frac{V^*}{R} = \frac{50\ EE\text{-}3}{4.17} = .012\ A$

$R_s = \frac{(.012)(4.17)}{5 - .012} = .010\ \Omega$

b) $I_s = 5 - .012 = 4.988\ A$

c) THE METER READS 50 EE-3 WHEN 5 AMPS FLOW, SO MULTIPLIER IS

$\frac{5}{50\ EE\text{-}3} = 100$

---

49 FROM EQN 13.89

$Z = 200$
$a = P$
$\phi = 30{,}000$ LINES $= 3\ EE\text{-}4$ wb
$N = 3600$

$E = (200)(3\ EE\text{-}4)\left(\frac{3600}{60}\right) = 3.6\ V$

---

50 REFER TO FIGURE 13.23

a) FROM EQN 13.98

$E = 110 - (90)(.05) = 105.5$

b) $I_f = \frac{V_{line}}{R_f} = \frac{110}{60} = 1.83$

$I_{line} = I_F + I_a = 1.83 + 90 = 91.83$

---

51 $P_{IN} = I_{line} V_{line} = (67)(240) = 16{,}080\ W$

$R_B = \frac{V_B}{I_A} = \frac{2}{3.35} = .597\ R$

$I_{a,LOADED} = 67 - I_f = 67 - 3.16 = 63.84$ AMPS

$P_{LOSS} = I_f V_f + I_a^2 R_a + I_a^2 R_B$

$= (3.16)(240) + (63.84)^2(.207) + (63.84)^2(.597)$

$= 4035.1$

$\eta = \frac{16080 - 4035.1}{16080} = .749$

$HP = (.749)(16080)W(1.341\ EE\text{-}3)\ HP/W$

$= 16.15\ HP$

---

52 a) $I_f = \frac{V_t}{R_f} = \frac{121}{134} = .903$

AT NO LOAD, $I_f = I_a$ AND $E = I_a(R_a + R_f)$

SO $E = .903(.31 + 134) = 121.28$

AT LOAD,

$I_f = \frac{V}{R_f} = \frac{110}{134} = .821$

SO $I_{LOADED} = I + I_f = 30 + .821 = 30.821$

$E = 110 + (30.821)(.31) = 119.55$

b) SINCE $E \propto (RPM)$ {SEE EQN 13.89}

$RPM = \left(\frac{119.555}{121.28}\right) 1775 = 1749.8\ RPM$

c) $P_{LOSSES} = I_a^2 R_a + I_f^2 R_f$

$= (30.821)^2 .31 + (.821)^2 (134) = 384.8\ W$

E-I-T HOMEWORK SOLUTIONS: A. C. ELECTRICITY

1 FOR A MULTI-POLAR ALTERNATOR, FROM EQN 14.3 WITH $\sin\omega t = 1$,

a) $V_{MAX} = 2\pi N \phi \frac{p}{2}\frac{n}{60}$

$N = (12)\text{coils}\,(20)\frac{\text{TURNS}}{\text{COIL}} = 240 \text{ TURNS}$

$A = (.167)(.10) = .0167\ M^2$

$\phi = BA = (1)T(.0167)m^2 = .0167\ wb$

$\frac{p}{2} = 4/2 = 2 \text{ poles}$

$n = 1800 \text{ RPM}$

SO, $V_{MAX} = 2\pi(240)(.0167)(2)\frac{1800}{60} = 1511\ V$

b) $HP = \frac{(1)\,KW\,(1.341)}{(.9)} = 1.49\ hp$

2 a) FROM EQN 14.4

$n = \frac{(2)(60)\,f}{p} = \frac{(2)(60)(60)}{24} = 300 \text{ RPM}$

b) $V_{EFF} = 2200V$

$V_{MAX} = \sqrt{2}\,(2200) = 3111.3\ V$

FROM PROBLEM #1,

$\phi = \frac{(2)(60)(V_{MAX})}{(2\pi)Npn}$

$= \frac{(2)(60)(3111.3)}{(2\pi)(20)(24)(300)} = .413\ wb$

3 FROM PAGE 14-2 THE NUMBER OF CYCLES PER ARMATURE REVOLUTION IS $\frac{p}{2} = \frac{6}{2} = 3$. THEREFORE, $\omega_{POTENTIAL} = 3\,\omega_{ARMATURE}$

THE ELECTRICAL PHASE DIFFERENCE IS $3(20°) = 60°$

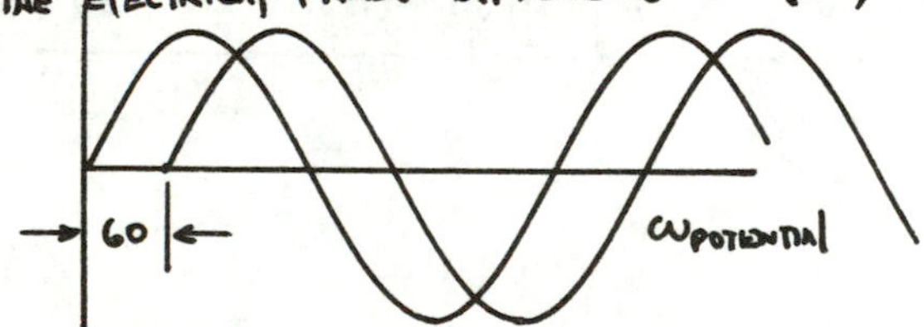

$V_{TERMINAL} = V[\sin(3\omega t) + \sin(3\omega t + 60)]$

WHERE ω IS THE ARMATURE SPEED

$\frac{dV}{dt} = V[\cos(3\omega t)(3\omega) + \cos(3\omega t + 60°)(3\omega)] = 0$

SO $\cos(3\omega t) + \cos(3\omega t + 60) = 0$

SO, EITHER $(3\omega t) + (3\omega t + 60) = 0$

OR $(3\omega t + 180) + (3\omega t + 60) = 0$

THE FIRST RESULTS IN $\omega t = -10°$ (NO GOOD). THE SECOND RESULTS IN $\omega t = -40°$ (OKAY)

PUTTING $\omega t = -40$ INTO THE $V_{TERMINAL}$ EQUATION,

$V_{TERMINAL} = V[\sin(-120) + \sin(-60)]$

$= -1.732\ V$

4 a) FROM EQN 14.16

$S_{ave} = \frac{1}{4}\int_0^4 S\,dt$

$= \frac{1}{4}\left[\int_0^2 S\,dt + \int_2^4 S\,dt\right]$

$= \frac{1}{4}\left[5t\Big|_0^2 + 3t\Big|_2^4\right]$

$= \frac{1}{4}[10 - 0 + 12 - 6] = \frac{16}{4} = 4$

NOTICE THAT THIS IS THE AREA UNDER THE CURVE DIVIDED BY THE PERIOD.

FROM EQN 14.19

$S_{RMS} = \sqrt{\frac{1}{4}\left((5)^2 t\Big|_0^2 + (3)^2 t\Big|_2^4\right)}$

$= \sqrt{\frac{1}{4}(50 - 0 + 36 - 18)} = 4.123$

$CF = \frac{5}{4.123} = 1.213$

$FF = \frac{4.123}{4} = 1.031$

b) $AREA = (\frac{1}{2})(3)(150) = 225$

$S_{ave} = \frac{225}{3} = 75$

NOW, $S = 50t$, SO

$S_{RMS} = \sqrt{\frac{1}{3}\int_0^3 (50t)^2\,dt}$

$= \sqrt{\frac{1}{3}\left(\frac{2500}{3}\right)t^3\Big|_0^3}$

$= \sqrt{\frac{2500}{9}(27 - 0)} = 86.60$

$CF = \frac{150}{86.6} = 1.732$

$FF = \frac{86.6}{75} = 1.155$

c) $AREA = (5)(80) = 400$

$S_{ave} = \frac{400}{20} = 20$

$S_{RMS} = \sqrt{\frac{1}{20}\int_0^5 (80)^2\,dt}$

$= \sqrt{\frac{1}{20}(6400)t\Big|_0^5} = \sqrt{\frac{1}{20}(6400)5} = 40$

$CF = \frac{80}{40} = 2$

$FF = \frac{40}{20} = 2$

d) $AREA = \frac{1}{2}(30)(7) = 105$

$S_{ave} = \frac{105}{10} = 10.5$

(MORE)

PROBLEM 4 CONTINUED

$$S = 30 - \frac{30}{7}(t)$$

$$S_{RMS} = \sqrt{\frac{1}{10}\left[\int_0^7 \left(30 - \frac{30}{7}t\right)^2 dt\right]}$$

$$= \sqrt{\frac{900}{10}\int_0^7 \left(1 - \frac{2}{7}t + \frac{t^2}{49}\right) dt}$$

$$= \sqrt{\frac{900}{10}\left[t - \frac{1}{7}t^2 + \frac{t^3}{3(49)}\right]_0^7}$$

$$= \sqrt{\frac{900}{10}\left(7 - 7 + \frac{7}{3}\right)} = 14.49$$

$$CF = \frac{30}{14.49} = 2.07$$

$$FF = \frac{14.49}{10.5} = 1.38$$

5

$$V_s = V_p\left(\frac{N_s}{N_p}\right) = 550\left(\frac{40}{200}\right) = 110V$$

$$I_s = \frac{V_s}{R} = \frac{110}{4.2} = 26.19 \text{ A}$$

$$I_p = 26.19\left(\frac{40}{200}\right) = 5.24 \text{ A}$$

6

$$\frac{1}{R_s} = \frac{1}{8} + \frac{1}{8} + \frac{1}{8}$$

$$R_s = 2.667$$

FROM EQN 14.62

$$a = \sqrt{Z_p/Z_s} = \sqrt{\frac{200}{2.667}} = 8.66$$

7

$$a = \frac{200}{100} = 2$$

FROM EQN 14.62

THE SOURCE SEES A LOAD OF

$$Z_t = \frac{V}{I_p} = \frac{8I_p + V_p}{I_p} = \frac{\frac{8}{2}I_s + 2V_s}{\frac{I_s}{2}}$$

$$= \frac{\frac{8}{2}I_s + 2(I_s)(6+J8)}{\frac{I_s}{2}} = 8 + 4(6+J8)$$

$$= 32 + J32$$

SO

$$|Z_t| = \sqrt{(32)^2 + (32)^2} = 45.25$$

AND

$$Z_t = 45.25\angle 45°$$

$$I_p = \frac{V}{Z} = \frac{100}{45.25\angle 45°} = 2.21\angle -45°$$

THE POWER FACTOR IS

$$\cos\left[\arctan\left(-8/6\right)\right] = .6$$

THE SECONDARY IMPEDANCE IS

$$|Z_s| = \sqrt{(6)^2 + (8)^2} = 10$$

THE SECONDARY CURRENT IS

$$I_s = (2)(I_p) = 2(2.21) = 4.42$$

ASSUME THE 100 VOLTS IS AN EFFECTIVE VALUE. THEN

$$P_s = I_s^2 Z_s \cos\theta = (4.42)^2(10)(.6) = 117.2 \text{ W}$$

8

a) $-15 + J0$

b) $10(\cos 37) = 7.99$

$10(\sin 37) = 6.02$

$7.99 + J6.02$

c) $50(\cos 120) = -25$

$50(\sin 120) = 43.3$

$-25 + J43.3$

d $0 - J21$

9

a) $\arctan\left(\frac{7}{6}\right) = 49.4$

$\sqrt{(7)^2 + (6)^2} = 9.22$

$9.22\angle 49.4$

b) $\arctan\left(\frac{-60}{50}\right) = -50.2$

$\sqrt{(60)^2 + (50)^2} = 78.1$

$78.1\angle -50.2$

c) $180 - \arctan\left(\frac{45}{-75}\right) = 149.06°$

$\sqrt{(45)^2 + (75)^2} = 87.46$

$87.46\angle 149.06$

d) $\arctan\left(\frac{-180}{90}\right) = -63.4$

$\sqrt{(180)^2 + (90)^2} = 201.2$

$201.2\angle -63.4$

10

a) $\frac{7}{6} + J\frac{5}{6} = 1.167 + J(.833)$

$\phi = \arctan\left(\frac{.833}{1.167}\right) = 35.52$

$\sqrt{(1.167)^2 + (.833)^2} = 1.43$

$1.43\angle 35.52$

(MORE)

b) $\frac{13+j17}{15-j10} = \frac{21.4\angle 52.59}{18.03\angle -33.69} = 1.19\angle 86.28$

c) $\frac{.020\angle 90}{.034\angle 56} = .588\angle 34$

11

$V_{C_1} = \frac{6}{3+6}(12) = 8V$

$V_{C_2} = \frac{3}{3+6}(12) = 4V$

12V, $C_1 = 3\mu f$, $C_2 = 6\mu f$

FROM EQN 14.30

$q_{C_1} = (3\text{ EE-}6)(8) = 24\text{ EE-}6\text{ C}$

$q_{C_2} = (6\text{ EE-}6)(4) = 24\text{ EE-}6\text{ C}$

12 a) $X_L = \omega L = 2\pi(1200)(350\text{ EE-}6) = 2.639\,\Omega$

b) $I = \frac{V}{X_L} = \frac{10}{2.639} = 3.789$

13

$Z_1 = 6.403\angle 51.34$

$Z_2 = 5.83\angle -30.96$

I_t, I_1, I_2; $4+j5$ (Z_1), $5-j3$ (Z_2)

$\frac{1}{Z_t} = \frac{1}{Z_1} + \frac{1}{Z_2} = \frac{1}{6.403\angle 51.34} + \frac{1}{5.83\angle -30.96}$

$= .1562\angle -51.34 + .1715\angle 30.96$

$= (.0976 - j.122) + (.1471 + j.0882)$

$= .2447 - j.0338$

$= .247\angle -7.86$

so $Z_t = \frac{1}{.247\angle -7.86} = 4.049\angle +7.86$

a) $I_t = \frac{V}{Z_t} = \frac{120}{4.04\angle 7.86} = 29.64\angle -7.86$

b) $I_1 = \frac{120}{6.403\angle 51.34} = 18.74\angle -51.34$

c) $I_2 = \frac{120}{5.83\angle -30.96} = 20.58\angle 30.96$

d) $P_{REAL} = \sum I^2R = (18.74)^2(4) + (20.58)^2 5$
$= 3522.4\text{ W}$

14 $Z_1 = 10\angle 0$

$Z_2 = 2\pi(60)(.02) = 7.54\angle 90$

$X_C = \frac{1}{2\pi(60)(1000\text{ EE-}6} = 2.653$

$Z_3 = 3 - j2.653 = 4\angle -41.49$

a) $I_1 = \frac{(100)V}{(10)\Omega} = 10\angle 0\text{ A}$

b) $I_2 = \frac{100}{7.54\angle 90} = 13.26\angle -90$

c) $I_3 = \frac{100}{4\angle -41.49} = 25\angle 41.49$

$I_t = 10\angle 0 + 13.26\angle -90 + 25\angle 41.49$

$= (10+j0) + (0 - j13.26) + (18.73 + j16.56)$

$= 28.73 + j3.3$

$= 28.92\angle 6.55$

$PF = \cos(6.55) = .9935$

15 $X_L = \omega L = (2000)(1) = 2000\,\Omega$

$X_C = \frac{1}{(2000)(.5\text{ EE-}6)} = 1000\,\Omega$

$Z_1 = 1000 - j1000 = 1414\angle -45$

$Z_2 = 3000 + j2000 = 3606\angle 33.69$

$Z_t = \frac{1}{\frac{1}{1414\angle -45} + \frac{1}{3606\angle 33.69}} + 500\angle 0$

$= \frac{1}{7.072\text{ EE-}4\angle 45 + 2.773\text{ EE-}4\angle -33.69} + 500\angle 0$

$= \frac{1}{5\text{EE-}4 + j5\text{EE-}4 + 2.307\text{EE-}4 - j1.538\text{ EE-}4} + 500\angle 0$

$= \frac{1}{7.307\text{ EE-}4 + j3.462\text{ EE-}4} + 500\angle 0$

$= \frac{1}{8.086\text{ EE-}4\angle 25.35} + 500\angle 0$

$= 1236\angle -25.35 + 500\angle 0$

$= 1117 - j529.2 + 500 + j0$

$= 1617 - j529.2$

$= 1701.4\angle -18.12$

$I_t = \frac{V}{Z} = \frac{141.4\angle 60}{1701.4\angle -18.12} = .0831\angle 78.12$

V ACROSS THE 2 PARALLEL LEGS IS

$141.4\angle 60 - 500(.0831\angle 78.12)$

$141.4\angle 60 - 41.55\angle 78.12$

$70.7 + j122.5 - (8.554 + j40.66)$

$= 62.15 + j81.84$

$= 102.8\angle 52.79$

(MORE)

PROBLEM 15 CONTINUED

$$I_{Z_1} = \frac{V}{Z} = \frac{102.8 \angle 52.79}{1414 \angle -45} = .0727 \angle 97.79$$

$$V_C = I X_C = (.0727) \angle 97.79 \, (1000 \angle -90)$$

$$= 72.7 \angle 7.79 \text{ VOLTS}$$

16

$$Z_t = 4 - J5 + \frac{1}{\frac{1}{7\angle -90} + \frac{1}{2\angle 90}}$$

$$= 4 - J5 + \frac{1}{.1429 \angle 90 + .5 \angle -90}$$

$$= 4 - J5 + \frac{1}{.3571 \angle -90}$$

$$= 4 - J5 + 2.8 \angle 90$$

$$= 4 - J5 + J2.8$$

$$= 4 - J2.2 = 4.565 \angle -28.81$$

$$I_t = \frac{V}{Z} = \frac{200 \angle 30}{4.565 \angle -28.81} = 43.81 \angle 58.81$$

$$V_{\text{ACROSS LEGS}} = 200 \angle 30 - I Z_{RC}$$

$$= 200 \angle 30 - (43.81 \angle 58.81)(4 - J5)$$

$$= 173.21 + J100 - (43.81 \angle 58.81)(6.40 \angle -51.34)$$

$$= 173.21 + J100 - 280.4 \angle 7.47$$

$$= 173.21 + J100 - (278 + J36.45)$$

$$= -104.79 + J63.55 = 122.6 \angle 148.8$$

17

a) $Z = \sqrt{(46)^2 + (11)^2} = 47.3 \angle -76.55$

b) $I = \frac{V}{Z} = \frac{120 \angle 0}{47.3 \angle -76.55}$

$= 2.537 \angle 76.55$

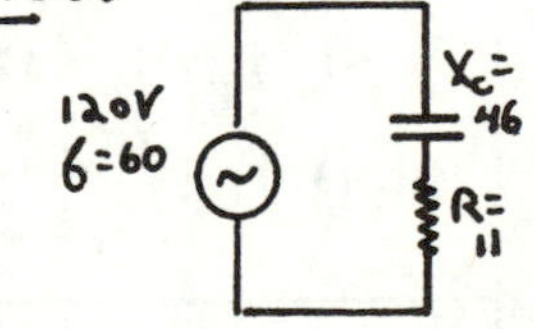

c) $X_C = X_L$ AT RESONANCE

$46 = (2\pi)(60)L$

$L = .122 h$

d) $V_C + V_L = 0$ AT RESONANCE

18

$X_C = X_L$

a) $\frac{1}{2\pi(60)C} = 20$

$C = 1.326 \text{ EE-4 F}$

b) SINCE THE RESISTANCE IS ZERO,

$I = \frac{V}{0} = \infty$

19

a) $X_C = X_L$

$$4 = \frac{1}{2\pi(60)C}$$

$C = 6.63 \text{ EE-4 F}$

b) FROM EQN 14.77,

$$Q = \frac{1}{2\pi(60)(6.63\text{EE-4})(4)} = 1.00$$

c) FROM EQN 14.83

$$\omega_2 - \omega_1 = \frac{2\pi(60)}{1.0} = 377 \text{ RAD/SEC}$$

20

$\omega^* = 2\pi f = 2\pi(100 \text{EE3}) = 6.28 \text{ EE5 RAD/SEC}$

FROM EQN 14.86,

$$C = \frac{1}{L\omega^{*2}} = \frac{1}{(100\text{EE-6})(6.28\text{EE5})^2}$$

$= 2.54 \text{ EE-8 F}$

21

FROM EQN 14.84,

$$\omega^* = \frac{1}{\sqrt{(1\text{EE-3})(2\text{EE-9})}} = 7.071 \text{ EE5}$$

$$\frac{R}{2L} = \frac{10}{2(1\text{EE-3})} = 5000$$

FROM EQN 14.82,

$\omega^* \pm \frac{R}{2L} = 712{,}100$ AND $702{,}100$ RAD/SEC

22

$P = I^2 R$

$1152 = (12)^2(6+R)$

$R = 2\,\Omega$

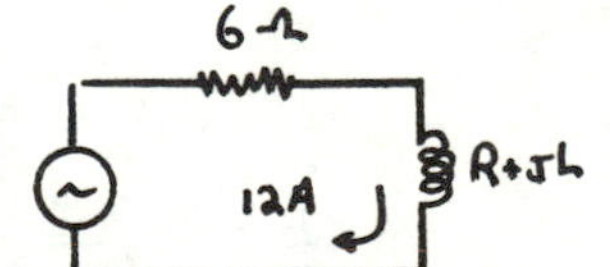

$X_L = 2\pi(60)L = 377L$

$$Z_{total} = \sqrt{(6+2)^2 + (377L)^2}$$

$$= \sqrt{64 + (377L)^2}$$

BUT $Z = \frac{V}{I} = \frac{120}{12} = 10$

$$10 = \sqrt{64 + (377L)^2}$$

$L = .0159 h$

23

BEFORE

$\phi_1 = \text{ARCCOS}(.6)$

$= 53.13$

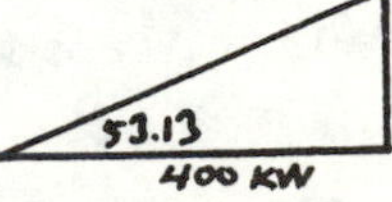

SO KVARS = 400(TAN 53.13) = 533.33

AFTER

$\phi_2 = \text{ARCCOS}(.8)$

$= 36.87$

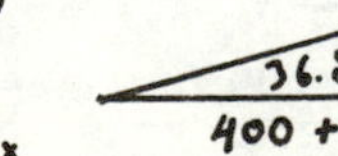

KVARS = (400+150) x TAN(36.87) = 412.5

SO THE MOTOR TAKES (533.33 − 412.5) = 120.83 KVAR (MORE)

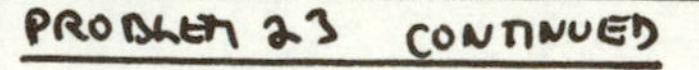

PROBLEM 23 CONTINUED

MOTOR

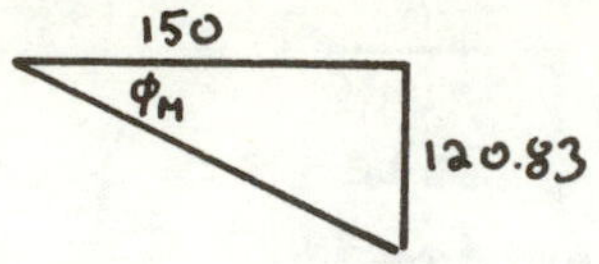

$$\text{POWER FACTOR} = \cos\phi_M = \cos\left(\text{ARCTAN}\ \frac{-120.83}{150}\right) = .779$$

24

BEFORE

$\phi_1 = \text{ARCCOS}(.82) = 34.92$

$\text{KW} = (10{,}000)(\cos 34.92) = 8200$

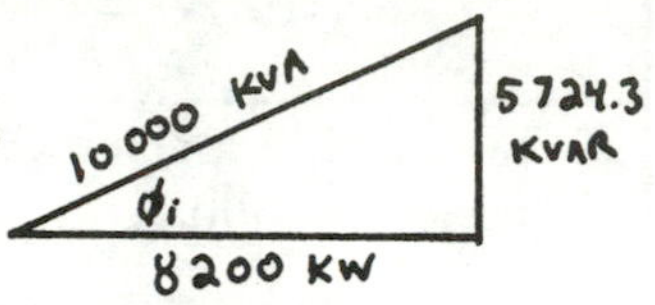

$\text{KVAR} = (10{,}000)(\sin 34.92) = 5724.3$

MOTORS

$$\text{KW} = \frac{(3)(2000)\ \text{HP}\ (.7457)\ \text{KW/HP}}{.92} = 4863.3\ \text{KW}$$

$\phi_M = \text{ARCCOS}(.8) = -36.87°$ LEADING

$\text{KVAR} = (\tan(36.87))\ 4863.3 = 3647.5$

AFTER

$$\phi_2 = \text{ARCTAN}\ \frac{2076.8}{13063.3} = 9.03$$

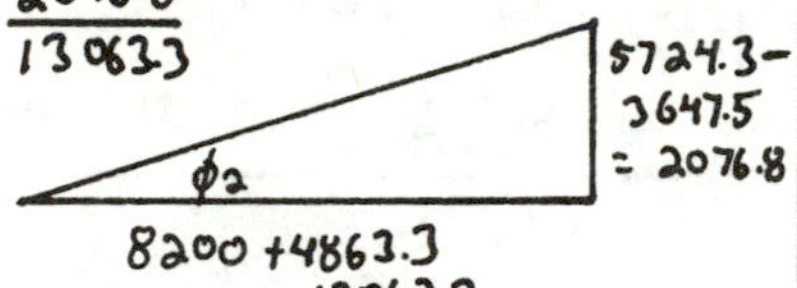

a) $\text{PF} = \cos\phi_2 = .988$

b) $\text{KVA} = \dfrac{2076.8}{\sin(9.03)} = 13232\ \text{KVA}$

25

$\text{VA} = 50_J$ SO

$I = \dfrac{50}{100} = .5\text{A}$

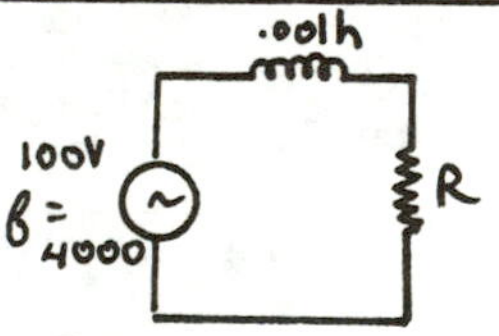

$X_L = 2\pi f L = 2\pi(4000)(.001) = 25.13\ \Omega$

$$I = \frac{V}{Z} \quad .5 = \frac{100}{\sqrt{(25.13)^2 + (R)^2}}$$

$R = 198.4$

26

$V_{LINE} = 208 \qquad \cos\phi = 1$

FROM EQN 14.116

$3000 = \sqrt{3}\ V_L I_L$

BUT FROM EQN'S 14.112

$$I_p = I_L = \frac{V_L}{\sqrt{3}R}$$

$$\text{SO } 3000 = \frac{\sqrt{3}\ V_L V_L}{\sqrt{3}\ R} = \frac{(208)^2}{R}$$

$R = 14.42\ \Omega$ EACH

27 FROM EQN 14.111

$5{,}000{,}000 = \sqrt{3}(4160)(I_{LINE})(.84)$

$I_{LINE} = 826.1\ \text{A}$

28 FROM EQN 14.116

a) $$P_{t,\text{WYE}} = \sqrt{3}\ V_L I_L = \sqrt{3}\ V_L\left(\frac{V_L}{\sqrt{3}R}\right) = \frac{V_L^2}{R} = \frac{(120)^2}{10} = 1440$$

b) FROM EQN 14.111

$P_{t,\text{delta}} = \sqrt{3}\ V_L I_L$

BUT $I_L = \dfrac{\sqrt{3}\ V_L}{R}$

$$P_t = \frac{3V_L^2}{R} = \frac{3(120)^2}{10} = 4320$$

29

a) $I_p = \dfrac{V}{Z} = \dfrac{208}{20\angle 25} = 10.4\angle{-25}$

b) FROM EQN 14.107

$I_L = \sqrt{3}\ I_p = \sqrt{3}(10.4) = 18.0\angle{-30-25} = 18\angle{-55}$

c) $V_p = V_L = 208\text{V}$

d) $P_p = I_p V_p \cos\phi = (10.4)(208)\cos(-25) = 1961\text{W}$

e) $P_t = 3P_p = 3(1961) = 5883$

30

$Z = 5\angle 53.13$

a) $V_p = \dfrac{V_L}{\sqrt{3}} = \dfrac{110}{\sqrt{3}} = 63.5$

b) $I_L = \dfrac{110}{\sqrt{3}\ 5\angle 53.13} = 12.7\angle{-53.13}$

c) $I_p = I_L = 12.7\angle{-53.13}$

d) $P_t = \sqrt{3}(110)(12.7)\cos(-53.13) = 1451.8$

31 FROM TABLE 14.3

$$q_C = (100\ \text{EE-6})(100)\left[1 - e^{\frac{-.05}{(1000)(100\ \text{EE-6})}}\right] = 3.935\ \text{EE-3 C}$$

FROM EQN 14.41

$$P_C = \frac{(.5)(3.935\ \text{EE-3})^2}{100\ \text{EE-6}} = .07741\ \text{J}$$

32

(a) SINCE THE CURRENT THROUGH THE INDUCTOR CANNOT CHANGE INSTANTLY,

$$I_L(t=0^+) = 0$$

(b) FROM TABLE 14.3

$$I = \frac{100}{80}\left[1 - e^{-\frac{80t}{7}}\right]$$

$$= 1.25\left[1 - e^{-11.43t}\right]$$

AT $t=2$,

$$I = 1.25\left[1 - e^{-(11.43)(2)}\right]$$

$$= 1.25 \text{ AMPS}$$

33

$$I_0 = \frac{V}{R} = \frac{100}{15} = 6.667$$

FROM TABLE 14.3,

$$I = 6.667\, e^{\frac{(-15)1}{8}} = 1.022$$

34

FROM TABLE 14.3, AT $t = 5EE\text{-}4$ Sec

$$I = \frac{100}{80}\left(1 - e^{\frac{-80(5EE-4)}{.15}}\right) = .2926 \text{ A}$$

AFTER THE SWITCH IS CLOSED TO POSITION 2, THE CIRCUIT WILL BOTH CHARGE AND DISCHARGE.

CHARGING

$$I = \frac{50}{80}\left(1 - e^{-\frac{80t}{.15}}\right) = .625\left(1 - e^{-533.3t}\right)$$

DISCHARGING

$$I = .2926\, e^{-\frac{80t}{.15}} = .2926\, e^{-533.3t}$$

$$I_{total} = .625 - .3324\, e^{-533.3t}$$

35

THIS IS A CASE OF SIMULTANEOUS CHARGING AND DISCHARGING.

$$V_0 = \frac{q}{C} = \frac{(600EE-6)\,C}{(200EE-6)\,F} = 3V$$

$$I = \left(\frac{60}{1000}\right)e^{\frac{-t}{(1000)(200EE-6)}} + \frac{3}{1000}\, e^{\frac{-t}{(1000)(200EE-6)}}$$

$$= .063\, e^{-5t}$$

WHEN $t=0$,

$$I = .063 \text{ A}$$

36

SINCE THE CURRENT CANNOT CHANGE INSTANTANEOUSLY, $I(0) = 0$

37

$$P = \eta IV\cos\phi$$

$$I = \frac{P}{\eta V\cos\phi} = \frac{(20)\,HP\,(745.7)\,W/HP}{(.86)(440)(.76)}$$

$$= 51.86 \text{ A} \quad \text{(single phase)}$$

38

FROM EQN 14.132,

$$N_S = \frac{(120)(60)}{4} = 1800 \text{ RPM}$$

a) FROM EQN 14.133,

$$N_r = (1-.03)1800 = 1746 \text{ RPM}$$

b)

$$\text{TORQUE} = \frac{(5252)\,HP}{RPM} = \frac{(5252)(200)}{1746}$$

$$= 601.6 \text{ FT-LBF}$$

c) FROM EQN 14.111,

$$\frac{(200)\,HP\,(745.7)\,W/HP}{.85} = \sqrt{3}\,(440)\,I_{LINE}\,(.91)$$

$$I_{LINE} = 253 \text{ A}$$

39

$$\phi = \text{ARC COS}(.82) = 34.92$$

$$KVAR = 550\,\text{TAN}(34.92) = 384$$

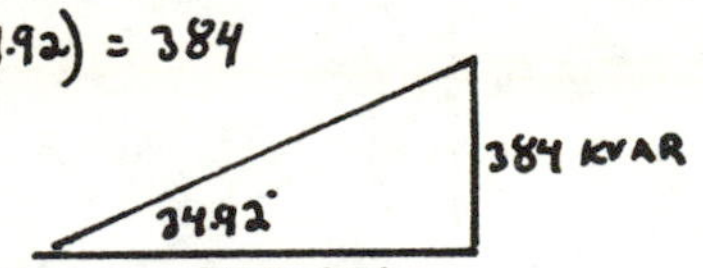

THE NEW MOTOR

$$KW = (250)\,HP \times (.7457)\,\frac{KW}{HP} = 186.4$$

AFTER

$$KVAR = (736.4)\,\text{TAN}(\text{ARCCOS}\ .95) = 242.0$$

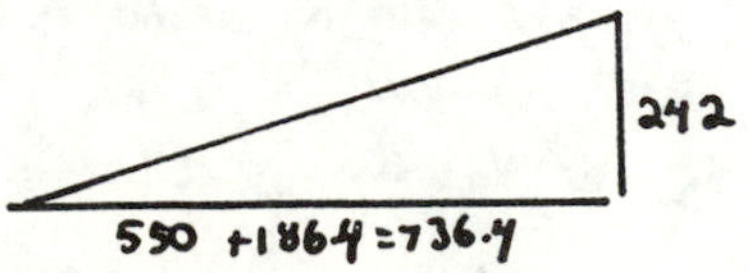

SO, THE CHANGE IN KVAR IS

$$(384 - 242) = 142 \text{ KVAR}$$

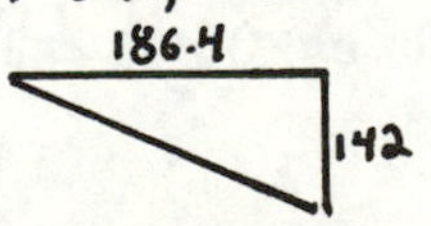

$$KVA = \sqrt{(142)^2 + (186.4)^2} = 234.3$$

40 ASSUME 4 POLES
FROM EQN 14.132

$$f = \frac{(960)(4)}{120} = 32$$

THAT'S NOT CLOSE TO ANYTHING IN COMMERCIAL USE, SO TRY 6 POLES

$$f = \frac{(960)(6)}{120} = 48 \text{ HZ}$$

SO, 50 HZ {EUROPEAN}

41 FROM EQN 14.139

a) $\mu = \frac{-\Delta V_b}{\Delta V_c}$ {I_b CONSTANT}

$$= \frac{-(200-100)}{-1-0} = 100$$

b) FROM EQN 14.141 {V_b CONSTANT}

$$g_m = \frac{\Delta I_b}{\Delta V_c} = \frac{(2-.5)\text{EE-3}}{0-(-1)} = 1.5 \text{ EE-3}$$

c) FROM EQN 14.137

$$r_p = \frac{\Delta V_b}{\Delta I_b} \approx \frac{100-0}{(2-0)\text{EE-3}} = 50{,}000\ \Omega$$

CHECK:

$$\mu = g_m r_b$$
$$100 = (1.5 \text{ EE-3})(50000)$$
$$100 = 75 \checkmark \text{ \{close enough\}}$$

42 THE LOAD LINE IS DEFINED BY

$$(V_b, i_b) = (400, 0) \text{ AND } \left(0, \frac{400}{133{,}000}\right) = 3 \text{ ma}$$

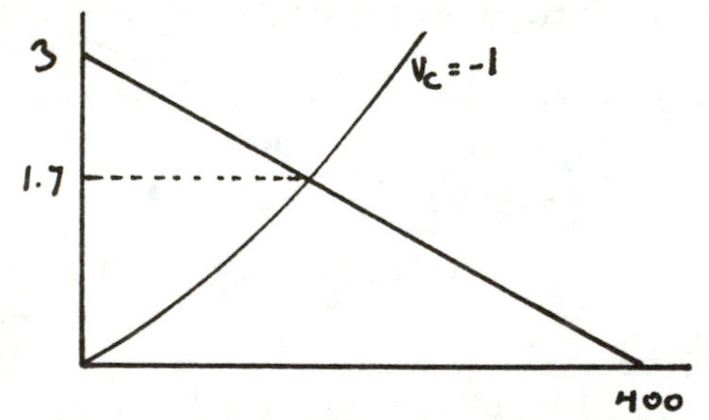

THE GRID VOLTAGE WITH A SIGNAL IS $(-10+9) = -1$

a) THE PLATE CURRENT IS 1.7 ma

b) AT $V_g = 8$, $V_b = -2$, $i_b \approx 1.2$ mA

AT $V_g = 10$, $V_b = 0$, $i_b \approx 2.3$ mA

$$\Delta i_b = 2.3 - 1.2 = 1.1 \text{ mA}$$

c) $\Delta V_L = \Delta i_b R_L = (1.1 \text{ EE-3})(133 \text{ EE3}) = 146.3 \text{ V}$

d) $\frac{\Delta V_L}{\Delta V_g} = \frac{146.3}{2} = 73$

43 FROM THE SENTENCE BELOW THE EQN 14.151

$$I_s \approx 70\ \mu A$$

a) FROM EQN 14.151

$$I = (70)\mu A\left[e^{40(.2)} - 1\right] = .209 \text{ A}$$

b) FROM EQN 14.150

$$I = (70)\mu A\left[e^{\frac{(1.6\text{EE-19})(.2)}{(1.38\text{ EE-23})(273+40)}} - 1\right]$$

$$= .115 \text{ A}$$

E-I-T HOMEWORK SOLUTIONS: MATERIALS SCIENCE

1 FROM FIGURE 10.20

$d_{1-2} = 4\sqrt{2/3}\ r$

BUT $a = \frac{4r}{\sqrt{3}}$

SO, $d_{1-2} = \frac{4\sqrt{2/3}\ r}{\frac{4r}{\sqrt{3}\,a}} = \sqrt{2}\ a$

SIMILARLY,

$d_{1-3} = \frac{4r}{\frac{4r}{\sqrt{3}\ a}} = \sqrt{3}\ a$

2 a) INTERCEPTS ARE (1,0,1) SO DIRECTION IS

[1,0,1]

b) INTERCEPTS ARE $(1, -\frac{1}{2}, 0)$
NO FRACTIONS ARE ALLOWED, SO ALL NUMBERS ARE MULTIPLIED BY 2 ⟶ (2,-1,0). THE OVERBAR IS USED FOR NEGATIVE NUMBERS, SO THE DIRECTION IS

$[2,\bar{1},0]$

c) THE INTERCEPTS ARE (0,1,0)

[0,1,0]

3 USE THE SAME COORDINATE SYSTEM AS IN PROBLEM #2. FOLLOW THE PROCEDURE ON PAGE 10-19

a) THE INTERCEPTS ARE $(1, \frac{1}{2}, \infty)$. TAKING THE RECIPROCALS ⟶ (120)

b) INTERCEPTS ARE $(1, \infty, \infty)$. TAKING THE RECIPROCALS ⟶ (100)

c) INTERCEPTS ARE (1,-1,1). TAKING THE RECIPROCALS ⟶ $(1\bar{1}1)$

4 a) 1 IN THE CENTER, $\frac{1}{8}$ AT EACH CORNER

$1 + 8(\frac{1}{8}) = 2$

b) $\frac{1}{8}$ AT EACH CORNER, $\frac{1}{2}$ ON EACH SURFACE

$8(\frac{1}{8}) + 6(\frac{1}{2}) = 4$

c) THE HEXAGONAL CELL IS $\frac{1}{3}$ OF THE STRUCTURE SHOWN IN FIGURE 10.18 #8. THIS IS SIMILAR TO BCC ⟶ 2 ATOMS/CELL

5 a) $\sin^2\phi = (\sin 13.7)^2 = .056$

$(\sin 16.0)^2 = .076$

$(\sin 22.9)^2 = .151$

$(\sin 27.1)^2 = .208$

$(\sin 28.3)^2 = .225$

$(\sin 32.6) = .290$

IF THE BASIC MULTIPLE IS ASSUMED TO BE .0187, THE RATIOS ARE 2.99, 4.06, 8.07, 11.1, 12.0, 15.5
SO, IT LOOKS LIKE FCC

b) IF $\sin^2(\theta_{min}) = .0187$

$\sin(\theta_{min}) = .137$

FROM EQN 10.19

$(1)(1.54)\text{Å}\,(\text{EE-}10)\,\text{m}/\text{Å} = (2)\,d_{hkl}\,(.137)$

$d_{hkl} = 5.62$ EE-10 M

6 FROM EQN 10.21

$(1)(.58\ \text{EE-}10)\ \text{m} = (2)\ d_{hkl}\ (\sin(9.5))$

$d_{hkl} = 1.76$ EE-10 M

7 a) THE DIAGONAL LENGTH AC IS

$\sqrt{(2R)^2 + (2R)^2} = 2.828R$

BUT THE DIAGONAL IS ALSO EQUAL TO $2r + 2R$

SO,

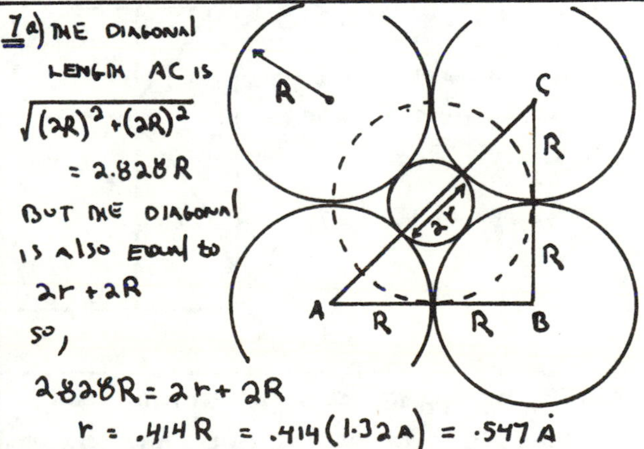

$2.828R = 2r + 2R$

$r = .414R = .414(1.32\text{Å}) = .547\text{Å}$

b) FROM TABLE 8.2, THE MINIMUM RATIO IS .732 FOR A COORDINATION NUMBER OF 8.

$r = (.732)(1.32) = .966\text{Å}$

8 PACKING FACTOR = $\frac{\text{VOLUME OF ATOMS}}{\text{VOLUME OF CELL}}$

a) FOR SIMPLE CUBIC (1 ATOM PER CELL)

$PF = \frac{\frac{4}{3}\pi r^3}{(2r)^3} = .527$

b) FOR FCC (4 ATOMS PER CELL)

$PF = \frac{4(\frac{4}{3}\pi r^3)}{(2\sqrt{2})^3 r^3} = .74$

c) FOR BCC (2 ATOMS PER CELL)

$PF = \frac{2(\frac{4}{3}\pi r^3)}{(4r/\sqrt{3})^3} = .68$

9 $J = -D\frac{dC}{dX}$

J = FLUX (ATOMS/CM^2-SEC)

D = DIFFUSION CONSTANT (CM^2/SEC)

$\frac{dC}{dX}$ = CONCENTRATION GRADIENT $\left(\frac{ATOMS}{CM^4}\right)$

10 $D = .4\ EE\text{-}8\ CM^2/SEC$

$$\frac{dC}{dX} = \frac{-.5\ \frac{ATOMS}{CM^3}}{5\ CM} = -.1\ \frac{ATOMS}{CM^4}$$

$$J = (.4\ EE\text{-}8)(-.1) = -(4\ EE\text{-}10)\ \frac{ATOMS}{CM^2\text{-}SEC}$$

TOTAL QUANTITY = 2J

$$= (2)CM^2(-4\ EE\text{-}10)\ \frac{ATOMS}{CM^2\text{-}SEC} = -8\ EE\text{-}10\ \frac{ATOMS}{SEC}$$

17 a)

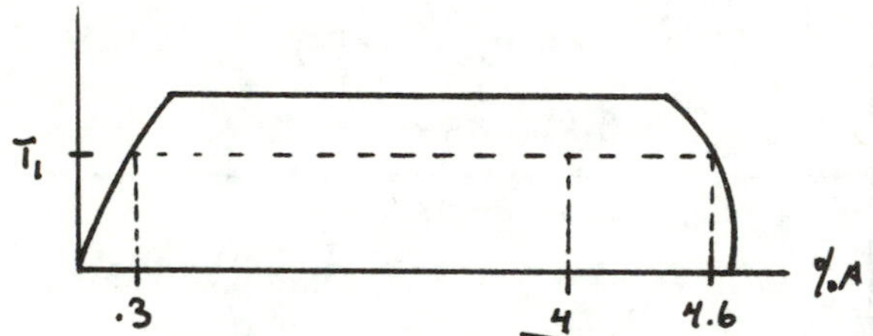

α IS .3% A, 99.7% B

β IS 4.6% A, 95.4% B

b) LINE SEGMENT XY = 9 MM

LINE SEGMENT XZ = 20 MM

$$\%\ LIQUID = \frac{9}{20} = 45\%$$

% SOLID = 1 − .45 = 55%

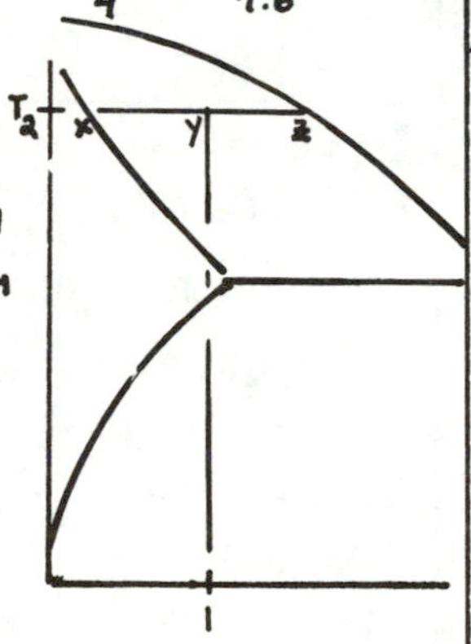

18 REFER TO FIGURE 10.15 AS THE ALLOY COOLS FROM 1650°F TO THE A_3 LINE, THE COMPOSITION IS 100% AUSTENITE WITH .5% CARBON. COOLING FROM THE A_3 LINE TO THE A_1 LINE, FERRITE PRECIPITATES FROM THE AUSTENITE. SINCE THERE IS VERY LITTLE CARBON IN FERRITE, THE CARBON CONCENTRATION INCREASES IN THE REMAINING AUSTENITE.

AT THE A_1 LINE, THE AMOUNT OF AUSTENITE IS

$$\frac{.5 - .025}{.8 - .025} = 61.3\%$$

UPON COOLING BELOW THE A_1 LINE, THE AUSTENITE TURNS INTO PEARLITE, SO THE % PEARLITE = 61.3%

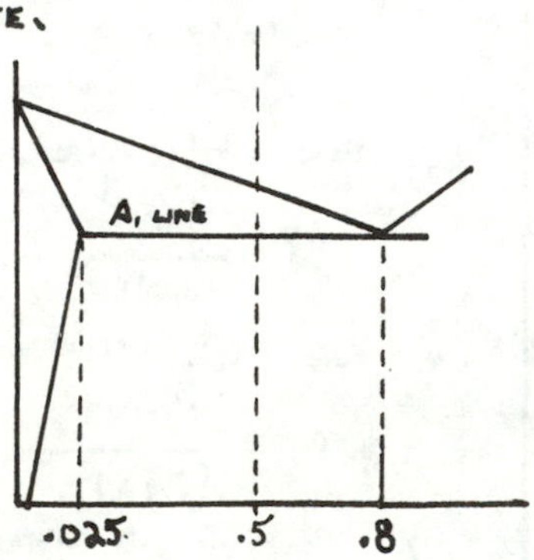

(.613)(50) = 30.65 LBM

ADDITIONAL CARBON PRECIPITATES OUT AS THE TEMPERATURE, BUT THE AMOUNT OF PEARLITE IS CONSTANT

19 1040

↳ .4% CARBON

(.004)(10) = .04 LBM CARBON

21 FROM FIGURE 10.15, THE A_2 LINE IS AT 1410°F OR 750°C

23 FROM A CHEMISTRY TEXTBOOK

a) −.771 V

b) +1.67 V

25 % CHANGE IN DIAMETER = (.3)(.020)

= .006

$$\sigma_{true} = \frac{P}{A_o(1-.006)^2} = 20242\ PSI$$

$$\epsilon_{true} = \ln\left(\frac{L}{L_o}\right) = \ln(1+.02) = .0198\ IN/IN$$

26 a)

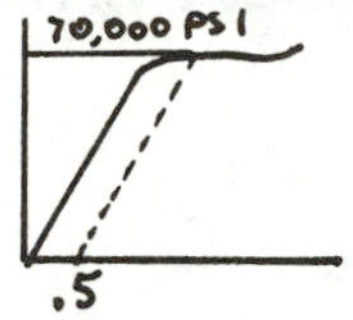

b)

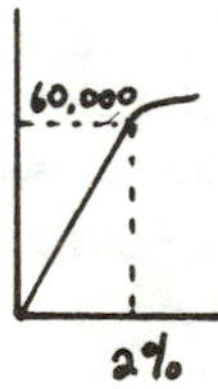

$$SLOPE = \frac{60{,}000}{.02} = 3\ EE6\ PSI$$

c) THE HIGHEST STRESS REACHED IS APPROXIMATELY 80 KSI

d) THE STRESS AT FRACTURE IS APPROXIMATELY 70 KSI

e) STRAIN AT FRACTURE ≈ 8% (6% AFTER SNAPPING BACK)

27 FROM EQN 11.77

$$G = \frac{E}{2(1+\mu)} = \frac{3\ EE6}{2(1+.3)} = 1.15\ EE6\ PSI$$

28 $$DUCTILITY = \frac{4 - 3.42}{4} = .145\ \text{REDUCTION IN AREA}$$

29 TOUGHNESS IS THE AREA UNDER THE CURVE: DIVIDE THE AREA INTO (20 KSI × 1%) SQUARES. THERE ARE ABOUT 25 SQUARES

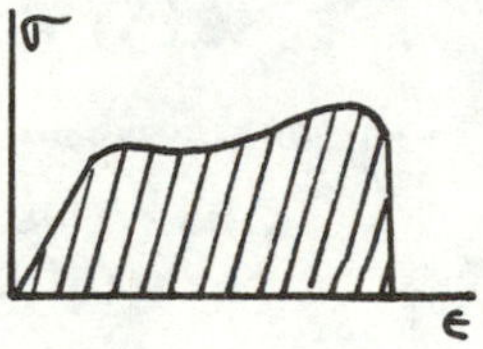

$$(25)(20{,}000)\ \frac{LBF}{IN^2}(.01)\frac{IN}{IN} = 5000\ \frac{IN\text{-}LBF}{IN^3}$$

30 PLOT THE DATA AND DRAW A STRAIGHT LINE

ε .063, .0175, 100 TIME

$$\text{slope} = \frac{.063-.0175}{100} = .000455\ 1/\text{HR}$$

33 THE VINYL CHLORIDE MER IS

H H
| |
C = C
| |
H Cl

THE MOLECULAR WEIGHT IS

$$2(12) + 3(1) + 1(35.5) = 62.5$$

WITH 20% EFFICIENCY, WE WILL NEED 5 MOLECULES OF HCl PER PVC MOLECULE TO SUPPLY THE END -Cl ATOM. THIS IS THE SAME AS 5 MOLES HCl PER MOLE PVC, OR,

$$\frac{(5)(6.023\ EE\ 23)}{7000} = (4.3\ EE\ 20)\ \frac{\text{molecules HCl}}{\text{gram PVC}}$$

34 MOLECULAR WEIGHTS: $H_2O_2 = 34$; $C_2H_4 = 28$.

THE NUMBER OF H_2O_2 MOLECULES IN 10 ML IS

$$\frac{(10)\text{ mL }(1)\text{ g/mL }(0.2/100)(6.023\ EE\ 23)\ \frac{\text{molecules}}{\text{gmole}}}{34\ \text{g/gmole}} = 3.54\ EE\ 20$$

THE NUMBER OF ETHYLENE MOLECULES IS

$$\frac{(12)(6.023\ EE\ 23)}{28} = 2.58\ EE\ 23$$

SINCE IT TAKES 1 H_2O_2 MOLECULE (THAT IS, 2 OH^- RADICALS) TO STABILIZE A POLYETHYLENE MOLECULE, THERE ARE 3.54 EE20 POLYMERS. THE DEGREE OF POLYMERIZATION IS

$$\frac{2.58\ EE23}{3.54\ EE20} = 729$$

44 USE THE SOLID VOLUME METHOD

(a)

MATERIAL	RATIO	WT PER SACK CEMENT	SOLID DENSITY		ABSOLUTE VOLUME (FT^3/SACK)
CEMENT	1.0	94	195	94/195 =	.48
SAND	1.9	179	165	179/165 =	1.08
COARSE	2.8	263	165	263/165 =	1.60
WATER				7/7.48 =	.94
					4.10

THEREFORE THE YIELD IS 4.10 FT^3 PER SACK OF CEMENT.

(b) THE NUMBER OF ONE-SACK BATCHES IS

$$\frac{(45)\ YD^3\ (27)\ FT^3/YD^3}{(4.1)\ FT^3/\text{SACK}} = 296.3\ \text{SACKS}\ \{\text{SAY } 297\}$$

$$\frac{(297)(1.9)(94)}{2000} = 26.5\ \text{TONS SAND}$$

$$\frac{(297)(2.8)(94)}{2000} = 39.1\ \text{TONS OF COARSE AGGR.}$$

$$(297)(7) = 2079\ \text{GALLONS OF WATER}$$

E-I-T HOMEWORK SOLUTIONS: PERIPHERAL SCIENCES

1 FROM EQN 15.11

$$\frac{I_1}{r_1^2} = \frac{I_2}{r_2^2}$$

SO $I_2 = \left(\frac{120}{80}\right)^2 20 = 45$ CANDELA

2 FROM TABLE 15.2

$\eta_L = 16.3$

FROM EQN 15.6

$F = P\eta_L = (100)\,W\,(16.3)\,\frac{\text{LUMENS}}{W} = 1630$ LUMENS

FROM EQN 15.10

$$E = \frac{F}{4\pi r^2} = \frac{1630}{4\pi(5)^2} = 5.19 \text{ LUMENS}/M^2$$

3 RADIATION PRESSURE $= \frac{\text{WATTS}}{M^2}$

$$= \frac{400}{4\pi(20)^2} = .0796 \text{ WATTS}/M^2$$

4 FROM EQN 15.3

$$\frac{f'}{f} = \frac{\lambda}{\lambda'} = \frac{\left(1-\frac{V}{c}\right)}{\sqrt{1-\left(\frac{V}{c}\right)^2}}$$

$$\frac{4350}{6580} = \frac{1-\frac{V}{3EE8}}{\sqrt{1-\left(\frac{V}{3EE8}\right)^2}}$$

BY TRIAL + ERROR,

$V = 1.17$ EE8 M/SEC

5 FROM EQN 15.2

$$f = \frac{c}{\lambda} = \frac{(3EE8)\ M/SEC}{(.22)\ M} = 1.364 \text{ EE9 HZ}$$

FROM EQN 15.5

$\Delta E = (6.626 \text{ EE-34})$ Joule-sec $(1.364 \text{ EE9})\ \frac{1}{\text{sec}}$

$= 9.035$ EE-25 J

6 FROM EQN 15.25

$$M = \frac{-3\ M}{1.5\ M} = -2$$

7 FROM EQN 15.36

$(1)(5 \text{ EE-5})$ CM $= \left(\frac{1}{2000}\right)\frac{\text{CM}}{\text{LINE}}(\sin\theta)$

$\sin\theta = .1$

(triangle: angle θ, base 3M, side Y)

$y = 3(\tan\theta) = .3$ M

8 FROM EQN 15.36

$(1)(5890)\,A\,(\text{EE-10})\frac{M}{A} = \left(\frac{1}{10{,}000}\right)\frac{\text{CM}}{\text{LINE}}\left(\frac{1}{100}\right)\frac{M}{\text{CM}}(\sin\theta)$

$\theta = \arcsin(.589) = 36.09°$

9 FROM EQN 15.24

a) $\frac{1}{3} + \frac{1}{OB} = \frac{1}{5}$

$OB = -7.5$

b) $\frac{1}{7} + \frac{1}{OB} = \frac{1}{5}$

$OB = 17.5$

c) $\frac{1}{12} + \frac{1}{OB} = \frac{1}{5}$

$OB = 8.57$

10 FROM PAGE 1-38

$C = 9.84$ EE8 FT/SEC

FROM EQN 15.15

$$n = \frac{(9.84 \text{ EE8})\ FT/SEC}{(5280)\frac{FT}{MI}\ (124{,}000)\ \frac{\text{miles}}{\text{sec}}} = 1.5$$

11 FROM EQN 15.13

$$\phi_{ci} = \arcsin\left(\frac{1}{1.70}\right) = 36.03°$$

12 FROM EQN 15.22

$$N = \frac{\sin\frac{1}{2}(60+45)}{\sin\frac{1}{2}(60)} = 1.587$$

13 ASSUME FUNDAMENTAL RESONANCE. FROM EQN 15.47

$\lambda = 4L = 4(4) = 16$ FT

FROM EQN 15.40

$V = f\lambda = (256)(16) = 4096$ FT/SEC

14 FROM EQN 15.44

$V \propto \sqrt{F}$ SO $F \propto V^2$

BUT $V = f\lambda$, SO

$F \propto f^2$

$$\frac{F_1}{F_2} = \left(\frac{f_1}{f_0}\right)^2 = \left(\frac{512}{512-2}\right)^2 = 1.0078$$

INCREASE ~ .0078

15 SINCE THE SOUND IS MINIMUM,

$10 - 8 = \frac{1}{2}\lambda$ OR $\lambda = 4$ FT

FROM EQN 15.40

$$f = \frac{V}{\lambda} = \frac{1100\ FT/SEC}{4} = 275 \text{ HZ}$$

16 FROM EQNS 15.44 AND 15.46

$$f = \frac{\sqrt{FL/M}}{2L} = \frac{1}{2}\sqrt{F/LM}$$

SO $F = 4f^2LM$

{MORE}

PROBLEM 16 CONTINUED

$F = 4(262)^2 \frac{1}{sec^2}(30)\,cm\,(30)\,g$

$= 2.47 \text{ EE8 } \frac{g\text{-}cm}{sec^2}$

$= 2.47 \text{ EE8 dynes}$

17 THE RATIO OF FREQUENCIES AN OCTAVE APART IS 2:1.

$f = 2 \times 2 \times 262 = 1048 \text{ Hz}$

18 FROM TABLE 15.8, $a = 1130$ FT/sec

FROM EQN 15.40

$\lambda = \frac{V}{f} = \frac{1130}{20} = 56.5 \text{ FT}$

19 a) FROM EQN 15.42,

$V = \sqrt{gE/\rho}$

FROM TABLE 11.1, $E = 10$ EE6 PSI

$V = \sqrt{\frac{(32.2)(10 \text{ EE6})(12)^2}{(.100)(12)^3}} = 16381 \text{ FT/sec}$

b) FROM EQN 15.43,

$a = \sqrt{(1.4)(32.2)(53.3)(460+120)}$

$= 1180.5 \text{ FT/sec}$

c) FROM PAGE 4-32 AT 50°;

$E = 305$ EE3 PSI AND $\rho = 62.41$ pcf

FROM EQN 15.42,

$V = \sqrt{\frac{(32.2)(305 \text{ EE3})(12)^2}{62.4}} = 4760 \text{ FPS}$

20 $f_2 = 2f_1 = 2(528) = 1056 \text{ Hz}$

$f_3 = 3f_1 = 3(528) = 1584 \text{ Hz}$

21 FROM EQN 15.46, (FOR $m=0$)

$V_{TRANS} = 2fL$

$= (2)(198)(.9)\,m = 356.4 \text{ m/sec}$

22 FROM EQN 15.52,

$\text{LOUDNESS} = 10 \log\left(\frac{5 \text{ EE-10}}{\text{EE-12}}\right) = 27 \text{ dB}$

23 FROM EQNS 15.44 AND 15.46,

$F = 4f^2LM$

$= 4(150)^2(4)\left(\frac{.038}{32.2}\right) = 424.8 \text{ LBF}$

24 FROM EQN 15.53,

$q = \frac{(2 \text{ EE-4})(1000)(170-50)}{4} = 6.0 \text{ cal/sec}$

25 FROM EQN 15.60,

$\frac{q}{A} = (.1713 \text{ EE-8})(1)\left[(1000)^4 - (530)^4\right]$

$= 1577.8 \text{ BTU/HR-FT}^2$

26 FOR AN ENCLOSED ISOTROPIC RADIATOR WITH $A_i \ll A_o$, $F_{1\text{-}2} = \epsilon_i$. FROM EQN 15.60,

$q = (.1713 \text{ EE-8})(.35)\left(\frac{5}{144}\right)\left[(5000)^4 - (530)^4\right]$

$= 1301 \text{ BTU/HR}$

FROM PAGE 1-36,

$(1301)\frac{\text{BTU}}{\text{HR}}(.2931)\frac{\text{W-HR}}{\text{BTU}} = 381.3 \text{ W}$

27 THE MAGNETIC AND ELECTRIC FIELDS ARE ORTHOGONAL, SO THE MAGNETIC FIELD IS EAST-WEST. WHEN THE MAGNETIC FIELD IS MAXIMUM, THE ELECTRIC FIELD IS ZERO, BUT THE ENERGY STORAGE IS THE SAME. FROM EQN'S 13.13 + 13.32

$\frac{1}{2}\epsilon E^2 = \frac{1}{2}\mu H^2$

$E = (\text{EE-3})\frac{V}{cm} = (\text{EE-1})\frac{V}{m}$

FROM PAGE 3-9,

$\epsilon = \frac{1}{(36\pi \text{ EE9})}$ AND $\mu = 4\pi \text{ EE-7}$

SO $H = E\sqrt{\frac{\epsilon}{\mu}} = (\text{EE-1})\sqrt{\frac{1}{(36\pi \text{ EE9})(4\pi \text{ EE-7})}}$

$= 2.65 \text{ EE-4 } \frac{\text{AMP}}{\text{METER}}$

28 $N_{Re} = \frac{Lv\rho}{\mu g} = \frac{Lv}{\nu}$

$N_{Pr} = \frac{c_p\mu}{k}$

$N_{Nu} = \frac{hL}{k}$

$N_{Bi} = \frac{hL}{k}$

E-I-T HOMEWORK SOLUTIONS: SYSTEMS THEORY

1

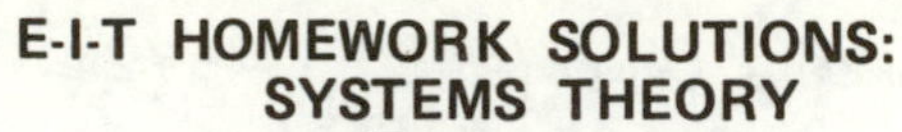

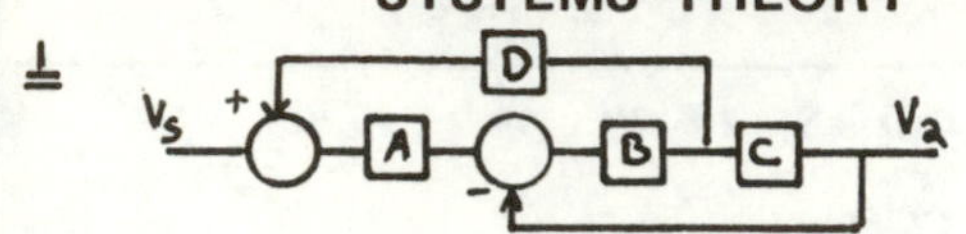

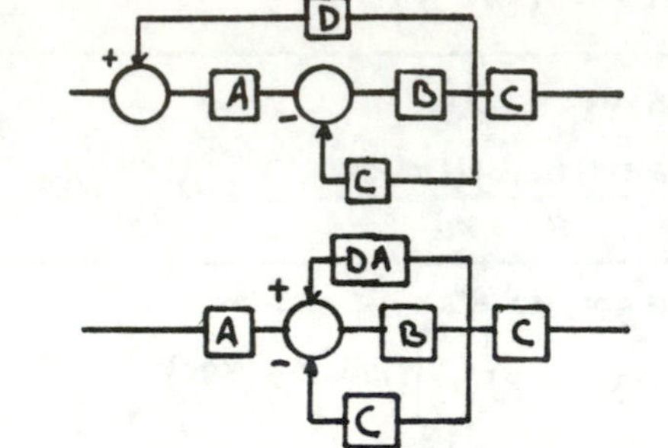

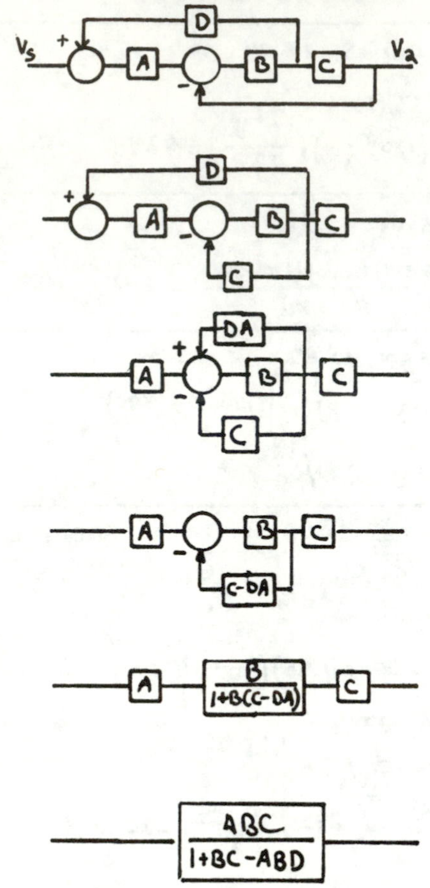

$A \; \left[\dfrac{B}{1+B(C-DA)}\right] \; C$

$$\frac{ABC}{1+BC-ABD}$$

2

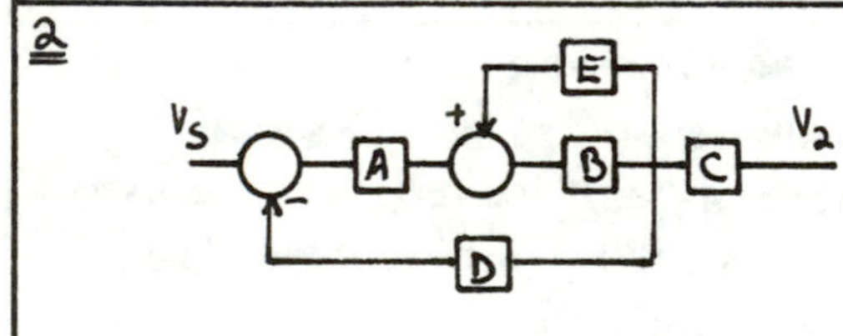

$A \; \left[\dfrac{B}{1-B(E-DA)}\right] \; C$

$$\frac{ABC}{1-BE+ABD}$$

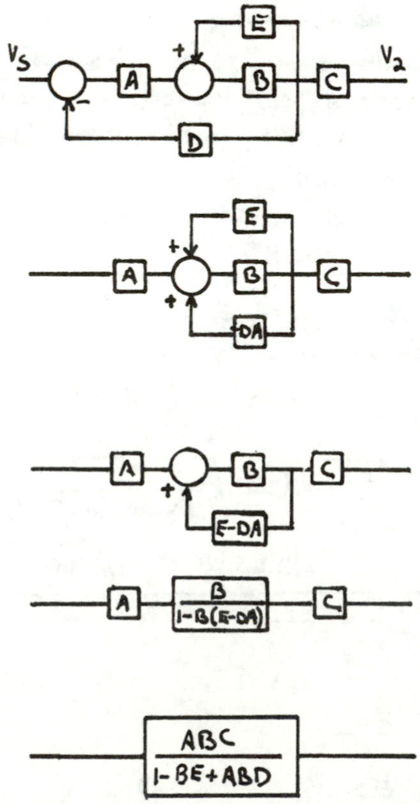

3.1

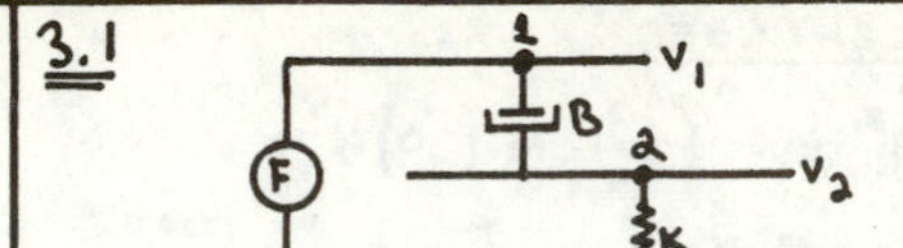

AT NODE 1

$F = B(V_1 - V_2)$ OR $F = B(x_1' - x_2')$

AT NODE 2

$F = K(x_2 - 0)$

3.2

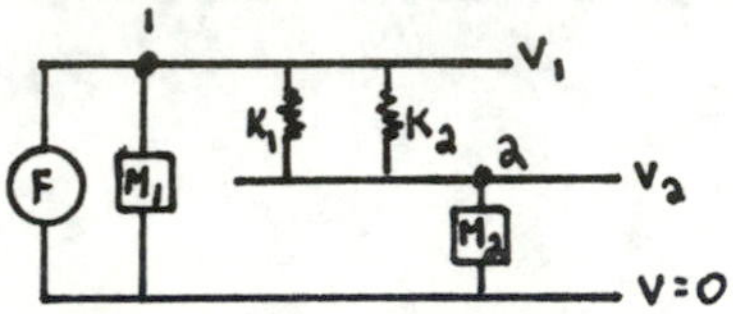

AT NODE 1

$F = M_1 a_1 + (K_1+K_2)(x_1 - x_2)$

$= M_1 x_1'' + (K_1+K_2)(x_1 - x_2)$

AT NODE 2

$0 = M_2 a_2 + (K_1+K_2)(x_2 - x_1)$

$= M_2 x_2'' + (K_1+K_2)(x_2 - x_1)$

3.3 DESPITE ITS APPEARANCES, THIS IS A ROTATIONAL SYSTEM.

τ = APPLIED ROTATIONAL TORQUE = FL

θ = ROTATED ANGLE

x_1 = ARC DISTANCE

x_2 = ARC DISTANCE

I = MOMENT OF INERTIA OF BEAM ABOUT AN END (FROM P. 9-20) $= \frac{1}{3} mL^2$

M = RESISTING MOMENT $= k x_2 l = k l \sin\theta \, l \approx k l^2 \theta$ for small values of θ

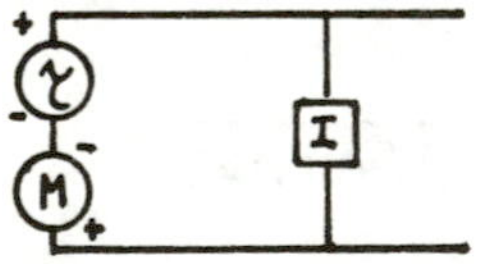

IN GENERAL, $\tau = I\alpha$, SO

$FL - kl^2\theta = \frac{1}{3} mL^2 \alpha$

BUT $\alpha = \theta''$

$$\theta'' + \left(\frac{3kl^2}{ML^2}\right)\theta = FL$$

3.4

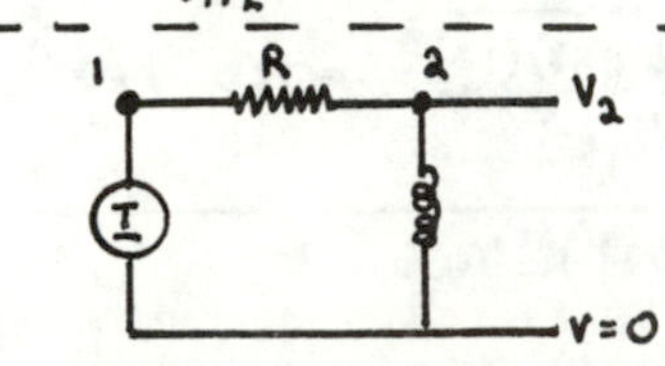

AT NODE 1

$I_1 = \frac{1}{R}(V_1 - V_2)$

AT NODE 2

$0 = \frac{1}{R}(V_2 - V_1) + \frac{1}{L}\int (V_2 - 0)\,dt$

OR, TAKING THE DERIVATIVE OF BOTH SIDES,

$0 = \frac{1}{R}(V_2' - V_1') + \left(\frac{1}{L}\right)V_2$

3.5

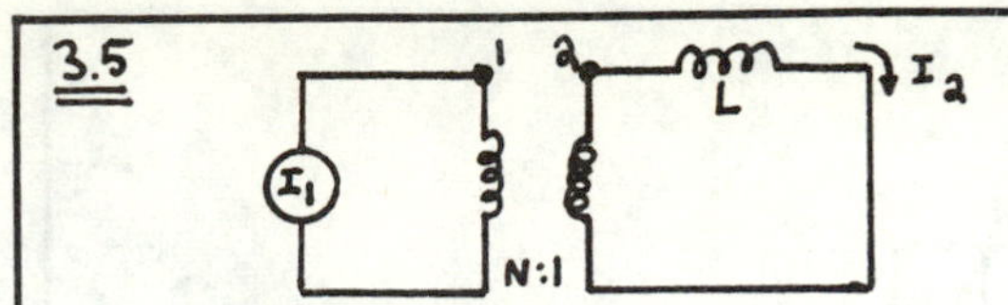

AT NODE 1

$$I_1 = I_1$$

AT NODE 2

$$I_2 = \frac{1}{L}\int V_2\,dt$$

BUT $V_2 = \frac{V_1}{N}$

SO $I_2 = \frac{1}{NL}\int V_1\,dt$

OR $I_2' = \frac{V_1}{NL}$

3.6

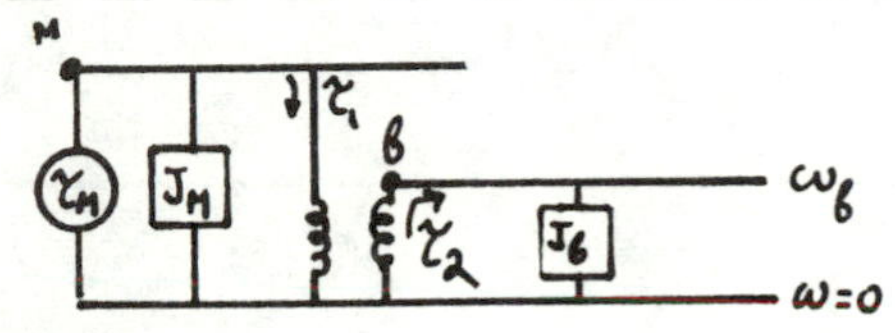

AT NODE M

$$\tau_M = J_M\alpha_M + \tau_1$$
$$= J_M\theta_M'' + \tau_1$$

AT NODE б

$$\tau_2 = J_б\alpha_б$$

BUT $\tau_2 = -\left(\frac{N_2}{N_1}\right)\tau_1$

SO $-\left(\frac{N_2}{N_1}\right)\tau_1 = J_б\,\theta_б''$

THE THIRD EQUATION REQUIRED IS

$$\theta_M = -\left(\frac{N_2}{N_1}\right)\theta_б$$

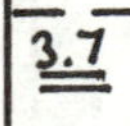

3.7

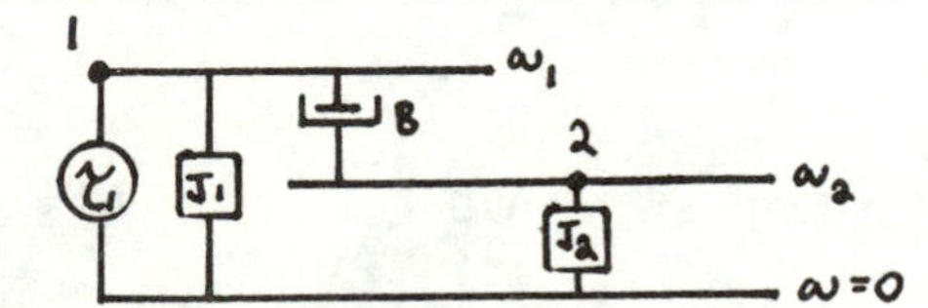

AT NODE 1

$$\tau_1 = J_1\alpha_1 + B(\omega_1 - \omega_2)$$
$$= J_1\theta_1'' + B(\theta_1' - \theta_2')$$

AT NODE 2

$$0 = J_2\alpha_2 + B(\omega_2 - \omega_1)$$
$$= J_2\theta_2'' + B(\theta_2' - \theta_1')$$

3.8

3.9

4 USE THE METHOD EXPLAINED ON PAGE 17-5

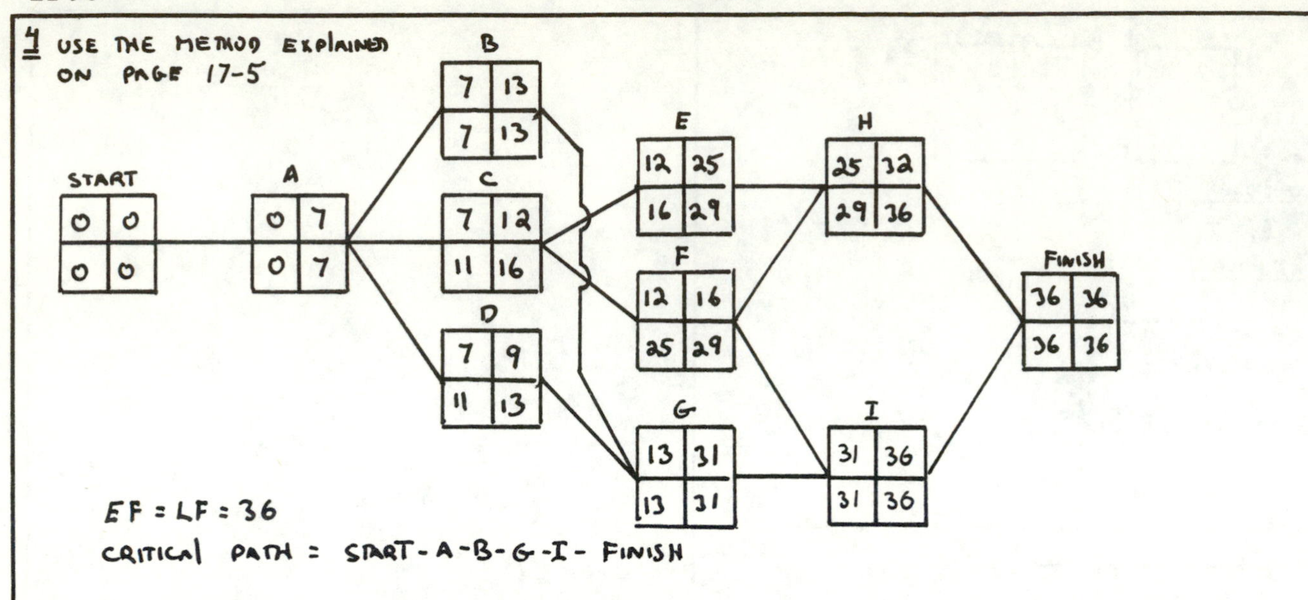

EF = LF = 36

CRITICAL PATH = START - A - B - G - I - FINISH

5

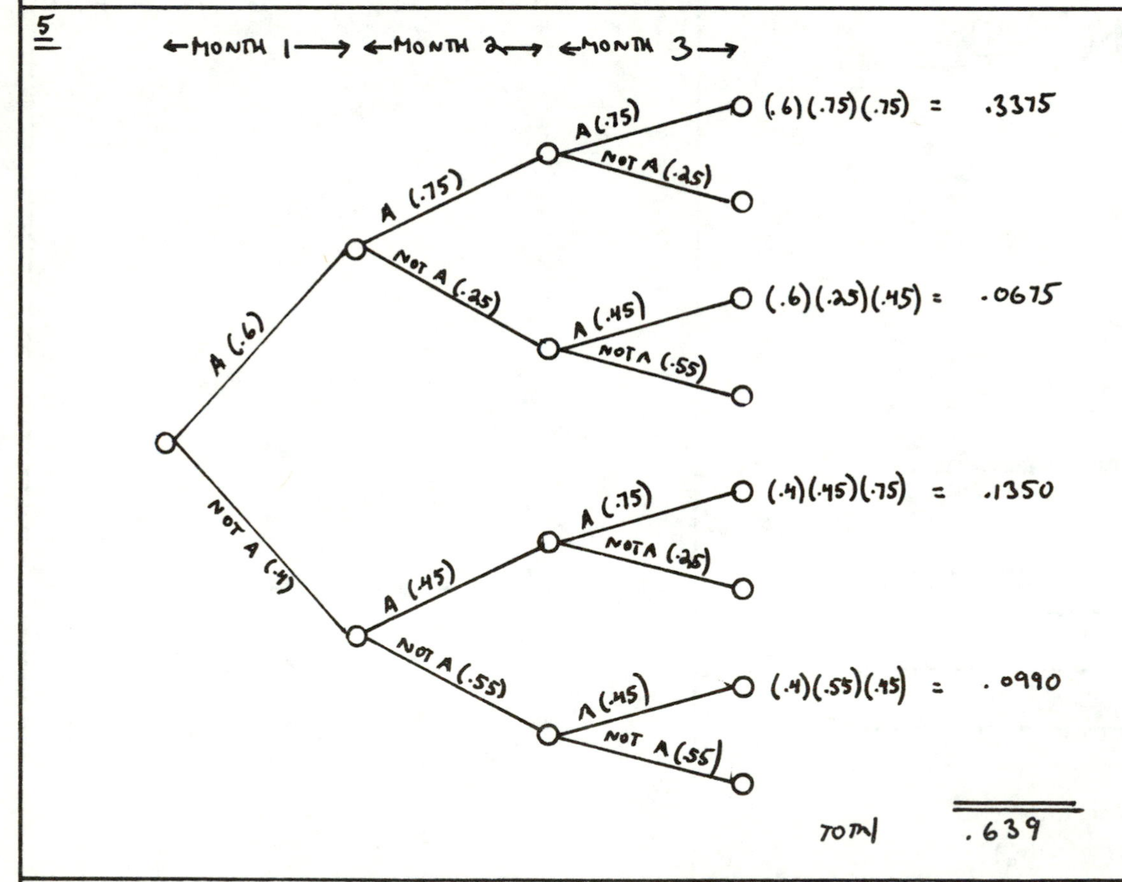

TOTAL .639

6 THE 1-STEP TRANSITION MATRIX IS

P = FROM	TO A	TO B
A	.75	.25
B	.45	.55

THE 2-STEP TRANSITION MATRIX IS

$$P^2 = \begin{bmatrix} .75 & .25 \\ .45 & .55 \end{bmatrix} \begin{bmatrix} .75 & .25 \\ .45 & .55 \end{bmatrix} = \begin{bmatrix} .675 & .325 \\ .585 & .415 \end{bmatrix}$$

FROM EQN 17.33

$$p_A^{(2)} = (.6)(.675) + (.4)(.585) = .639$$

NOTICE THAT THE 2-STEP MATRIX WAS NEEDED BECAUSE IN 3 MONTHS, ONLY 2 SWITCHES ARE MADE.

7 a)

$$MTBF = \frac{2(50) + 5(100) + 18(150) + 22(200) + 2(250) + 1(300)}{50}$$

$= 170$ HRS

b) REFER TO THE EXAMPLE TABLE ON PAGE 17-15

ELAPSED TIME	FAILURES	SURVIVORS	FAILURE PROB.	COND. DIST.
0	0	50	0	–
50	2	48	.04	.04
100	5	43	.10	.104
150	18	25	.36	.419
200	22	3	.44	.88
250	2	1	.04	.667
300	1	0	.02	1.00

8 FIRST FIND THE PROBABILITY THAT EACH BANK WILL BE OPERATIONAL - USE THE BINOMIAL DISTRIBUTION

BANK B

$p(0,2)$ = PROBABILITY THAT NONE OF THE 2 ELEMENTS WILL WORK

$$= \frac{2!}{(2-0)!\,0!}(.85)^0(.15)^2 = .0225$$

$R_B = 1 - .0225 = .9775$

BANK C

$$p(0,3) = \frac{3!}{0!(3-0)!}(.60)^0(.4)^3 = .064$$

$R_C = 1 - .064 = .936$

BANK D

THIS BANK WILL FAIL IF 2 OR 3 ELEMENTS FAIL.

$$p(0,3) = \frac{3!}{0!(3-0)!}(.8)^0(.2)^3 = .008$$

$$p(1,3) = \frac{3!}{1!(3-1)!}(.8)^1(.2)^2 = .096$$

$R_D = 1 - .008 - .096 = .896$

$$R^*_{system} = (.97)(.9775)(.936)(.896) = .795$$

9 163 OBSERVATIONS OF OPERATION A WERE MADE. THE PROBABILITY THAT A RANDOM SAMPLE OF OPERATION A WILL TAKE .25 MINUTES IS

$$\frac{3}{163} = .018$$

SIMILARLY, THE REST OF THE PROBABILITIES ARE FOUND, AS WELL AS THE CUMULATIVE DISTRIBUTION $F(t)$

t	p(t)	F(t)	1000 F(t)
.25	.018	.018	18
.30	.012	.030	30
.35	.061	.091	91
.40	.135	.226	226
.45	.166	.392	392
.50	.160	.552	552
.55	.147	.699	699
.60	.098	.797	797
.65	.092	.889	889
.70	.037	.926	926
.75	.031	.957	957
.80	.031	.988	988
.85	.012	1.000	1000

IT IS ARBITRARILY DECIDED THAT THE 1ST 3 NUMBERS IN EACH ROW IN TABLE 17.2 BE USED TO GENERATE RANDOM OPERATION TIMES. ONLY 10 SAMPLES WILL BE TAKEN ALTHOUGH MANY MORE WOULD NORMALLY BE TAKEN IN A REAL SIMULATION

SAMPLE	RANDOM #	TIME	CUM TIME
1	097	.40	0.40
2	542	.50	0.90
3	422	.50	1.40
4	019	.30	1.70
5	807	.65	2.35
6	065	.35	2.70
7	060	.35	3.05
8	269	.45	3.50
9	573	.55	4.05
10	796	.60	4.65

SIMILARLY, FOR THE OPERATION B:

t	p(t)	1000 F(t)
.25	.017	17
.30	.085	102
.35	.145	247
.40	.171	418
.45	.154	572
.50	.128	700
.55	.085	785
.60	.051	836
.65	.043	879
.70	.043	922
.75	.034	956
.80	.026	982
.85	.017	999

PROBLEM #9 CONTINUED

THE FIRST 3 NUMBERS OF EACH ROW STARTING WITH THE 5TH DOUBLE-DIGIT COLUMN ARE USED TO GENERATE 10 RANDOM SAMPLES:

sample	#	time
1	765	.55
2	648	.50
3	196	.35
4	093	.30
5	801	.60
6	340	.40
7	455	.45
8	020	.30
9	053	.30
10	035	.30

(hatched box) B IS IDLE; NO PART COMING

(hatched box) B IS IDLE; PART COMING ON CONVEYOR

NOW, A TIME LINE CAN BE DRAWN. THE FIRST PART TAKES 0.10 TO TRAVEL ON THE CONVEYOR BETWEEN A AND B.

TIME (MIN.): .4, .9, 1, 1.4, 1.7, 2, 2.35, 2.7, 3, 3.05, 3.5, 4, 4.05, 4.65, 5

A: 1 | 2 | 3 | 4 | 5 | 6 | 7 | 8 | 9 | 10

B: (idle) | 1 | 2 | 3 | 4 | (idle) | 5 | 6 | 7 | 8 | 9 | (idle) | 10

B time marks: .5, 1.05, 1.55, 1.9, 2.2, 2.45, 3.05, 3.45, 3.9, 4.2, 4.75, 5.05

THERE IS ESSENTIALLY NO INVENTORY IN FRONT OF OPERATION B.

OPERATION B PUT OUT 10 PIECES IN 5.05 MINUTES, SO OUTPUT $= \frac{10}{5.05} = 1.98$ PIECES/MIN

10 ASSUME A M/M/1 SYSTEM.

THE AVERAGE PROCESSING TIME OF OPERATION A IS

$$\frac{1}{163}\Big[3(.25)+2(.3)+10(.35)+22(.40)+27(.45)+26(.5)+24(.55)+16(.6)+15(.65)+6(.7)+5(.75)+5(.8)+2(.85)\Big]$$

$= .521$ MINUTES/ITEM

SO, THE MEAN ARRIVAL RATE TO OPERATION B IS $\lambda = \frac{1}{.521} = 1.919$

THE AVERAGE PROCESSING TIME FOR OPERATION B IS

$$\frac{1}{117}\Big[2(.25)+10(.3)+17(.35)+20(.4)+18(.45)+15(.5)+10(.55)+6(.6)+5(.65)+5(.7)+4(.75)+3(.8)+2(.85)\Big]$$

$= .479$ MINUTES/ITEM

SO, THE MEAN SERVICE TIME IS

$$\mu = \frac{1}{.479} = 2.088$$

FROM EQN 17.49, THE AVERAGE INVENTORY IS $\rho = \frac{\lambda}{\mu} = \frac{1.919}{2.088} = .919$

$$L_q = \frac{\rho\lambda}{\mu-\lambda} = \frac{(.919)(1.919)}{2.088-1.919} = 10.4$$

THE AVERAGE OUTPUT IS

$\mu = 2.088$ ITEMS PER MINUTE

11 ASSUME A M/M/1 SYSTEM.

$\lambda = 4/\text{HR}$

$\mu = \frac{60 \text{ MIN/HR}}{6 \text{ MIN/JOB}} = 10$ JOBS/HR

$\rho = \frac{\lambda}{\mu} = \frac{4}{10} = .4$

a) FROM EQN 17.44,

$p(0) = 1 - .4 = .60$ (60% IDLE)

b) IF SHE IS WORKING ON 1 JOB AND THERE ARE 3 WAITING TO BE STARTED, THERE ARE 4 IN THE SYSTEM.

$p(4) = (.6)(.4)^4 = .0154$

c) THE AVERAGE NUMBER OF JOBS WAITING TO BE STARTED IS GIVEN BY EQN 17.49:

{MORE}

PROBLEM 11 CONTINUED

$$L_q = \frac{(.4)(4)}{10-4} = .267$$

d) THE SERVICE DISTRIBUTION IS EXPONENTIAL, THAT IS, EQN 17.14

$$f(t) = \mu e^{-\mu t}$$

THE CUMULATIVE DISTRIBUTION IS

$$F(t) = \int_0^t f(t) = 1 - e^{-\mu t}$$

$$p\{\text{JOB TAKING LESS THAN } \tfrac{1}{2} \text{ HR}\}$$

$$= 1 - e^{-(10)(.5)} = .993$$

SO, THE PROBABILITY OF A JOB TAKING MORE THAN ½ HOUR IS 1 − .993 = .007

12 CHOOSE THE $ML\theta t$ SYSTEM. THEN

$$f = f(D, V, N, g, \rho, \mu)$$

$$\frac{ML}{\theta} = C_1 (L)^a \left(\frac{L}{\theta}\right)^b \left(\frac{1}{\theta}\right)^c \left(\frac{L}{\theta^2}\right)^d \left(\frac{M}{L^3}\right)^e \left(\frac{L^2}{\theta}\right)^f$$

THE NECESSARY EQUATIONS ARE

$$M: \quad 1 = e$$
$$L: \quad 1 = a + b + d + 2f - 3e$$
$$\theta: \quad 1 = b + c + 2d + f$$

13 a) THE EXPECTED PROFIT IS

$$(.25)(15) + (.30)(1) + (.45)(-6) = 1.35 \text{ million}$$

b) THE LOSS MATRIX IS

	OUTCOME θ_1	θ_2	θ_3
a_1 (INVEST)	−15	−1	6
a_2 (DON'T INVEST)	0	0	0

THE MAXIMUM LOSSES ARE

ACTION	MAX LOSS
a_1	6
a_2	0

THE MINIMAX CRITERION WOULD DIRECT YOU NOT TO INVEST SINCE THAT ACTION MINIMIZES YOUR MAXIMUM LOSS.

IF THE BAYE'S PRINCIPLE IS USED, THE EXPECTED LOSSES ARE:

$$a_1: (.25)(-15) + (.30)(-1) + (.45)(6) = -1.35$$

$$a_2: 0$$

MINIMIZING THE EXPECTED LOSS WOULD DIRECT YOU TO INVEST.

14

PROBLEM #14 CONTINUED

THERE ARE 2 POSSIBLE SOLUTIONS THAT MEET ALL CONSTRAINTS:

$(X_A, X_B) = (45, 55) \quad Z = 425$

$(80, 20) \quad Z = 600$

SO (80,20) IS THE OPTIMUM SOLUTION

15 LET X_A AND X_B BE THE NUMBER OF BARRELS OF FORMULAS A AND B RESPECTIVELY. THE OBJECT IS TO MAXIMIZE PROFIT.

MAX: $Z = 7X_A + 2X_B$

THE CONSTRAINTS ARE

DEMAND: $X_A + X_B = 100$

RESOURCES: $X_B \leq 55$

FISH: $2X_A + X_B \leq 180$

THIS IS A LINEAR, 2-DIMENSIONAL SYSTEM, WHICH MAY BE SOLVED GRAPHICALLY.

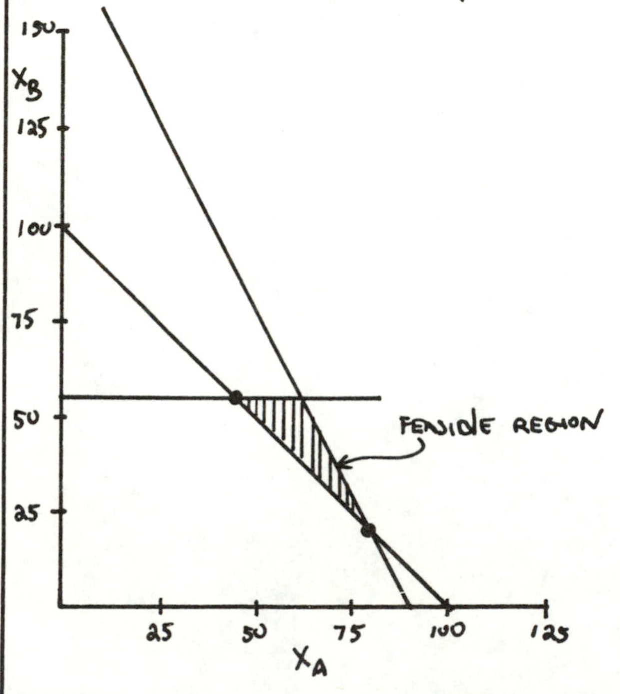

16 WRITE THE PROBLEM IN STANDARD FORM AND USE THE SIMPLEX METHOD

$$Z - 12X_1 - 18X_2 = 0$$
$$2X_1 + X_2 \leq 4$$
$$X_1 + 2X_2 \leq 4$$

NEXT, ADD SLACK VARIABLES

$$Z - 12X_1 - 18X_2 = 0$$
$$2X_1 + X_2 + X_3 = 4$$
$$X_1 + 2X_2 + X_4 = 4$$

THE INITIAL BASIS IS $X_1 = 0$, $X_2 = 0$, $X_3 = 4$, $X_4 = 4$, $Z = 0$

SINCE COEFFICIENTS IN THE FIRST ROW ARE NEGATIVE, WE CAN INCREASE Z BY INCREASING X_1 AND X_2. LET'S INCREASE X_2 BECAUSE IT HAS THE LARGEST NEGATIVE COEFFICIENT. X_2 CANNOT EXCEED 2 (FROM THE SECOND CONSTRAINT). IF $X_2 = 2$ THEN X_1 AND X_4 MUST BOTH BE 0. USING ROW OPERATIONS TO ELIMINATE ALL X_2'S EXCEPT THE X_2 IN THE 3RD ROW,

$$Z - 3X_1 + 9X_4 = 36$$
$$-3X_1 - 2X_3 + X_4 = -4$$
$$X_1 + 2X_2 + X_4 = 4$$

SO, $X_1 = 0$, $X_2 = 2$, $X_3 = 2$, $X_4 = 0$, $Z = 36$

SINCE X_1 STILL HAS A NEGATIVE COEFFICIENT, INCREASING IT WILL INCREASE Z. X_1 CANNOT INCREASE BEYOND $\frac{4}{3}$ (1ST CONSTRAINT). ELIMINATING ALL X_1'S EXCEPT THE X_1 IN THE 2ND ROW,

$$Z + 2X_3 + 8X_4 = 40$$
$$-3X_1 - 2X_3 + X_4 = -4$$
$$+2X_2 + \frac{2}{3}X_3 + \frac{2}{3}X_4 = \frac{8}{3}$$

SO $X_1^* = \frac{4}{3}$, $X_2^* = \frac{4}{3}$, $X_3^* = 0$, $X_4^* = 0$

$Z^* = 40$

17 TO MEET THE 2 CONSTRAINTS, ALL x_i MUST BE LESS THAN OR EQUAL TO ZERO, BUT x_1 AND x_2 MUST NOT BE NEGATIVE,

SO $x_1^* = 0$
$x_2^* = 0$

IF x_3 AND x_5 ARE STRICTLY LESS THAN ZERO, THEN Z WILL BE DECREASED,

SO $x_3^* = 0$
$x_5^* = 0$

HOWEVER, THE COEFFICIENT FOR x_4 IN THE OBJECTIVE FUNCTION IS NEGATIVE, SO DECREASING x_4 WITHOUT BOUND INCREASES Z WITHOUT BOUND

18a $y'' + 3y' + 2y = 0$

$$\mathcal{L}(y'') + 3\mathcal{L}(y') + 2\mathcal{L}(y) = 0$$

$$s^2\mathcal{L}(y) - s(1) + 3s\mathcal{L}(y) - 3(1) + 2\mathcal{L}(y) = 0$$

$$\mathcal{L}(y)\left[s^2+3s+2\right] = 3+s$$

$$\mathcal{L}(y) = \frac{3+s}{s^2+3s+2} = \frac{3}{(s+2)(s+1)} + \frac{s}{(s+2)(s+1)}$$

$$y = 3\left[\frac{1}{-2-(-1)}\left(e^{-2t} - e^{-t}\right)\right] + \frac{1}{2-1}\left[-2e^{-2t} + e^{-t}\right]$$

$$= -3e^{-2t} + 2e^{-2t} + 3e^{-t} - e^{-t}$$

$$= -e^{-2t} + 2e^{-t}$$

18b $y'' + 4y' + 4y = 0$

$$\mathcal{L}(y'') + 4\mathcal{L}(y') + 4\mathcal{L}(y) = 0$$

$$s^2\mathcal{L}(y) - s - 1 + 4s\mathcal{L}(y) - 4 + 4\mathcal{L}(y) = 0$$

$$\mathcal{L}(y)\left[s^2+4s+4\right] = 5+s$$

$$\mathcal{L}(y) = \frac{5+s}{\left[s^2+4s+4\right]} = \frac{5}{(s+2)(s+2)} + \frac{s}{(s+2)(s+2)}$$

USING PARTIAL FRACTIONS TO EXPAND THE SECOND TERM,

$$\frac{s}{(s+2)(s+2)} = \frac{A_1}{s+2} + \frac{A_2}{(s+2)^2}$$

$$= \frac{A_1(s+2) + A_2}{(s+2)^2} = \frac{A_1 s}{(s+2)^2} + \frac{A_2 + 2A_1}{(s+2)^2}$$

SO, $A_1 = 1$, $A_2 = -2$

$$\text{SO } \mathcal{L}(y) = \frac{5}{(s+2)^2} + \frac{1}{(s+2)} - \frac{2}{(s+2)^2}$$

$$= \frac{3}{(s+2)^2} + \frac{1}{(s+2)}$$

$$y = 3te^{-2t} + e^{-2t}$$

18c $y'' + 2y' + 2y = \cos(t)$

$$\mathcal{L}(y'') + 2\mathcal{L}(y') + 2\mathcal{L}(y) = \mathcal{L}(\cos(t))$$

$$s^2\mathcal{L}(y) - s + 2s\mathcal{L}(y) - 2 + 2\mathcal{L}(y) = \frac{s}{s^2+1}$$

$$\mathcal{L}(y)\left[s^2+2s+2\right] - s - 2 = \frac{s}{s^2+1}$$

$$\mathcal{L}(y)\left[s^2+2s+2\right] = \frac{s}{s^2+1} + s + 2$$

$$\mathcal{L}(y) = \frac{s}{(s^2+1)(s^2+2s+2)} + \frac{s}{s^2+2s+2} + \frac{2}{s^2+2s+2}$$

USING PARTIAL FRACTIONS ON THE FIRST TERM,

$$\frac{s}{(s^2+1)(s^2+2s+2)} = \frac{A_1 s + B_1}{(s^2+1)} + \frac{A_2 s + B_2}{s^2+2s+2}$$

WORKING WITH THE NUMERATORS ONLY,

$$s = (A_1 s + B_1)(s^2+2s+2) + (A_2 s + B_2)(s^2+1)$$

$$= s^3(A_1+A_2) + s^2(2A_1+B_1+B_2) + s(2A_1+2B_1+A_2) + 2B_1 + B_2$$

SO

$$\begin{aligned} A_1 + A_2 &= 0 \\ 2A_1 + B_1 + B_2 &= 0 \\ 2A_1 + A_2 + 2B_1 &= 1 \\ 2B_1 + B_2 &= 0 \end{aligned}$$

THIS HAS A SOLUTION OF

$A_1 = 1/5$ $A_2 = -\frac{1}{5}$ $B_1 = \frac{2}{5}$ $B_2 = -\frac{4}{5}$

$$\text{SO } \mathcal{L}(y) = \frac{\frac{1}{5}s}{s^2+1} + \frac{\frac{2}{5}}{s^2+1} - \frac{\frac{1}{5}s}{s^2+2s+2} - \frac{\frac{4}{5}}{s^2+2s+2} + \frac{2}{s^2+2s+2}$$

$$= \frac{\frac{1}{5}s}{s^2+1} + \frac{\frac{2}{5}}{s^2+1} + \frac{4}{5}\left[\frac{s+1}{(s+1)^2+1}\right] + \frac{\frac{2}{5}}{(s+1)^2+1}$$

$$y = \frac{1}{5}\cos t + \frac{2}{5}\sin t + \frac{4}{5}\left(e^{-t}\cos t\right) + \frac{2}{5}e^{-t}\sin t$$

18d $y'' + 2y' + 2y = e^{-t}$

$$\mathcal{L}(y'') + 2\mathcal{L}(y') + 2\mathcal{L}(y) = \mathcal{L}(e^{-t})$$

$$s^2\mathcal{L}(y) - 1 + 2s\mathcal{L}(y) + 2\mathcal{L}(y) = \frac{1}{s+1}$$

$$\mathcal{L}(y)\left[s^2+2s+2\right] = \frac{1}{s+1} + 1 = \frac{s+2}{s+1}$$

$$\mathcal{L}(y) = \frac{s+2}{(s+1)\left[s^2+2s+2\right]}$$

USING PARTIAL FRACTIONS,

$$\mathcal{L}(y) = \frac{A_1}{s+1} + \frac{A_2 s + B_2}{s^2+2s+2}$$

WORKING WITH THE NUMERATORS,

$$s+2 = A_1(s^2+2s+2) + (A_2 s + B_2)(s+1)$$

$$= s^2(A_1+A_2) + s(2A_1+A_2+B_2) + (2A_1+B_2)$$

(MORE)

PROBLEM 18d CONTINUED

so

$$A_1 + A_2 = 0$$
$$2A_1 + A_2 + B_2 = 1$$
$$2A_1 + B_2 = 2$$

so $A_1 = 1,\ A_2 = -1,\ B_2 = 0$

so $\mathcal{L}(y) = \frac{1}{s+1} - \frac{s}{s^2+2s+2}$

$$= \frac{1}{s+1} - \left[\frac{s+1}{(s+1)^2+1}\right] + \frac{1}{(s+1)^2+1}$$

so $y = e^{-t} - e^{-t}\cos t + e^{-t}\sin t$

18e $y'' + y = 1 - u_3$

$$\mathcal{L}(y'') + \mathcal{L}(y) = \mathcal{L}(1) - \mathcal{L}(u_3)$$
$$s^2\mathcal{L}(y) - 1 + \mathcal{L}(y) = \frac{1}{s} - \frac{1}{s}e^{-3s}$$
$$\mathcal{L}(y)[s^2+1] - 1 = e^{-3s}\left(-\frac{1}{s}\right) + \frac{1}{s}$$
$$\mathcal{L}(y) = e^{-3s}\left[\frac{-1}{s(s^2+1)}\right] + \frac{1}{s(s^2+1)}$$

USING PARTIAL FRACTIONS ON THE BRACKETED QUANTITY,

$$\frac{1}{s(s^2+1)} = \frac{A_1}{s} + \frac{A_2 s + B_2}{s^2+1}$$

WHICH HAS SOLUTIONS $A_1 = 1,\ A_2 = -1,\ B_2 = 0$

so $\mathcal{L}(y) = e^{-3s}\left[\frac{-1}{s} + \frac{s}{s^2+1}\right] + \frac{1}{s} - \frac{s}{s^2+1}$

$$y = u_3[-1 + \cos(t-3)] + 1 - \cos(t)$$

18f $y'' + 2y' + y = 1 - u_1$

$$\mathcal{L}(y'') + 2\mathcal{L}(y') + \mathcal{L}(y) = \mathcal{L}(1) - \mathcal{L}(u_1)$$
$$s^2\mathcal{L}(y) - s + 2s\mathcal{L}(y) - 2 + \mathcal{L}(y) = \frac{1}{s} - \frac{e^{-s}}{s}$$
$$\mathcal{L}(y)[s^2+2s+1] - s - 2 = \frac{1}{s} - \frac{e^{-s}}{s}$$
$$\mathcal{L}(y) = \frac{1}{s(s^2+2s+1)} + \frac{s}{s^2+2s+1} + \frac{2}{s^2+2s+1} - \frac{e^{-s}}{s(s^2+2s+1)}$$

USING PARTIAL FRACTIONS ON THE FIRST TERM,

$$\frac{1}{s(s^2+2s+1)} = \frac{A_1}{s} + \frac{A_2 s + B_2}{s^2+2s+1}$$

WORKING WITH THE NUMERATORS ONLY,

$$1 = A_1(s^2+2s+1) + (A_2 s + B_2)s$$

$$1 = s^2(A_1 + A_2) + s(2A_1 + B_2) + A_1$$

so $A_1 = 1$, $A_2 = -1$, $B_2 = -2$

so $\mathcal{L}(y) = \frac{1}{s} - \frac{2+s}{s^2+2s+1} + \frac{s}{s^2+2s+1} + \frac{2}{s^2+2s+1} - \frac{e^{-s}}{s(s^2+2s+1)}$

$$= \frac{1}{s} - e^{-s}\left[\frac{1}{s} - \frac{2}{s^2+2s+1} - \frac{s}{s^2+2s+1}\right]$$
$$= \frac{1}{s} - e^{-s}\left(\frac{1}{s} - \frac{2}{(s+1)^2} - \frac{s}{(s+1)^2}\right)$$

so $y = 1 - u_1(1 - 2\sin(t-1) - \cos(t-1))$

18g $y'' + 3y' + 2y = u_2$

$$\mathcal{L}(y'') + 3\mathcal{L}(y') + 2\mathcal{L}(y) = \mathcal{L}(u_2)$$
$$s^2\mathcal{L}(y) - 1 + 3s\mathcal{L}(y) + 2\mathcal{L}(y) = \frac{e^{-2s}}{s}$$
$$\mathcal{L}(y)[s^2+3s+2] = \frac{e^{-2s}}{s} + 1$$

AND $s^2+3s+2 = (s+2)(s+1)$

so $\mathcal{L}(y) = e^{-2s}\left[\frac{1}{s(s+2)(s+1)}\right] + \frac{1}{(s+2)(s+1)}$

so $y = u_2\left[\frac{1}{2}(1 + e^{-2(t-2)}) - 2e^{-(t-2)}\right] + e^{-t} - e^{-2t}$

18h $y'' + 2y' + 2y = P_4$

$$\mathcal{L}(y'') + 2\mathcal{L}(y') + 2\mathcal{L}(y) = \mathcal{L}(P_4)$$
$$s^2\mathcal{L}(y) - s + 2s\mathcal{L}(y) + 2 + 2\mathcal{L}(y) = e^{-4s}$$
$$\mathcal{L}(y)[s^2+2s+2] - s + 2 = e^{-4s}$$
$$\mathcal{L}(y) = e^{-4s}\left[\frac{1}{s^2+2s+2}\right] + \frac{s}{s^2+2s+2} - \frac{2}{s^2+2s+2}$$
$$= e^{-4s}\left[\frac{1}{(s+1)^2+1}\right] + \frac{s+1}{(s+1)^2+1} - \frac{3}{(s+1)^2+1}$$
$$y = u_4\left[e^{-(t-4)}\cos(t-4)\right] + e^{-t}\cos t - 3e^{-t}\sin t$$

18i $y'' + y = 2P_1$

$$\mathcal{L}(y'') + \mathcal{L}(y) = 2\mathcal{L}(P_1)$$
$$s^2\mathcal{L}(y) - s + \mathcal{L}(y) = 2\mathcal{L}(P_1)$$

{MORE}

PROBLEM 18 i CONTINUED

$$\mathcal{L}(y)\left[s^2+1\right] = 2e^{-s} + s$$

$$\mathcal{L}(y) = e^{-s}\left(\frac{2}{s^2+1}\right) + \frac{s}{s^2+1}$$

$$y = u_1\left(2\sin(t-1)\right) + \cos t$$

19a BY INSPECTION, $a_0 = 0$.

BECAUSE THIS FUNCTION HAS BOTH ODD AND HALF-WAVE SYMMETRY, ALL $a_N = 0$ AND $b_{N,\text{even}} = 0$

$$b_1 = \frac{1}{\pi}\int_0^{2\pi} f(t)\sin t\, dt$$

$$= \frac{1}{\pi}\int_0^{\pi}\sin(t) - \frac{1}{\pi}\int_{\pi}^{2\pi}\sin(t)$$

$$= \frac{1}{\pi}\left[-\cos(t)\right]_0^{\pi} + \frac{1}{\pi}\left[\cos(t)\right]_{\pi}^{2\pi}$$

$$= -\frac{1}{\pi}(-1-1) + \frac{1}{\pi}(1-(-1)) = \frac{4}{\pi}$$

IN GENERAL, $b_N = \frac{4}{\pi N}$ (N ODD)

SO $f(t) = \frac{4}{\pi}\left(\sin t + \frac{1}{3}\sin 3t + \frac{1}{5}\sin 5t + \cdots\right)$

19b BY INSPECTION, $a_0 = 0$. FROM EVEN FUNCTION SYMMETRY, ALL $b_N = 0$. FROM HALF-WAVE SYMMETRY, ALL $a_{N,\text{even}} = 0$.
FROM QUARTER-WAVE SYMMETRY,

$$\int_0^{2\pi} = 4\int_0^{\pi/2}$$

$$a_1 = \frac{4}{\pi}\int_0^{\frac{\pi}{2}} f(t)\cos t\, dt = \frac{4}{\pi}\int_0^{\frac{\pi}{2}}\cos t$$

$$= \frac{4}{\pi}\left[\sin t\right]_0^{\pi/2} = \frac{4}{\pi}(1-0) = \frac{4}{\pi}$$

IN GENERAL, $a_N = \frac{4}{\pi N}$ (N ODD)

SO $f(t) = \frac{4}{\pi}\left(\cos t + \frac{1}{3}\cos 3t + \frac{1}{5}\cos 5t + \cdots\right)$

19c BY INSPECTION, $a_0 = 0$. EVEN FUNCTION SYMMETRY, SO ALL $b_N = 0$. WITH HALF-WAVE SYMMETRY, $a_{N,\text{even}} = 0$. FINALLY, BECAUSE OF QUARTER-WAVE SYMMETRY,

$$\int_0^{2\pi} = 4\int_0^{\pi/2}$$

NOW, $f(x) = 1 - 2x$ BETWEEN 0 AND .5

TRANSFORMING THE X AXIS INTO THE t AXIS,

$$f(t) = 1 - \frac{2t}{\pi}$$

$$a_1 = \frac{4}{\pi}\int_0^{\frac{\pi}{2}}\left(1 - \frac{2t}{\pi}\right)\cos t\, dt$$

$$= \frac{4}{\pi}\left[\int_0^{\pi/2}\cos t - \frac{2}{\pi}\int_0^{\pi/2} t\cos t\right]$$

$$= \frac{4}{\pi}\left[\sin t\Big|_0^{\pi/2} - \frac{2}{\pi}(\cos t + t\sin t)\Big|_0^{\pi/2}\right]$$

$$= \frac{4}{\pi}\left[(1-0) - \frac{2}{\pi}\left(0 + \frac{\pi}{2} - 1 - 0\right)\right]$$

$$= 0.81$$

$$a_3 = \frac{4}{\pi}\int_0^{\pi/2}\left(1 - \frac{2t}{\pi}\right)\cos 3t$$

$$= \frac{4}{\pi}\left[\int_0^{\pi/2}\cos 3t - \frac{2}{\pi}\int_0^{\frac{\pi}{2}} t\cos 3t\right]$$

FROM INTEGRAL TABLES,

$$= \frac{4}{\pi}\left[\frac{1}{3}\sin 3t\Big|_0^{\pi/2} - \frac{2}{\pi}\left(\frac{\cos 3t}{9} + \frac{t\sin 3t}{3}\right)\Big|_0^{\pi/2}\right]$$

$$= \frac{4}{\pi}\left[\left(\frac{1}{3}\right)(-1) - 0 - \frac{2}{3\pi}\left(0 + \left(\frac{\pi}{2}\right)(-1) - \frac{1}{9} - 0\right)\right]$$

$$= \frac{4}{\pi}\left[-\frac{1}{3} + 0.357\right] = 0.03$$

SO, $f(t) \approx .81\sin t + 0.03\sin 3t + \cdots$

19d BY INSPECTION, $a_0 = \frac{1}{3}$

$f(t) = 10x \quad 0 \le x \le 1$

$f(t) = \frac{1}{2\pi}t \quad 0 \le t \le 2\pi$

$$a_1 = \frac{1}{\pi}\int_0^{2\pi} f(t)\cos t$$

$$= \frac{1}{\pi}\left[\int_0^{2\pi}\frac{1}{2\pi}(t)\cos t\right] = \frac{1}{2(\pi)^2}\int_0^{2\pi} t\cos t$$

{MORE}

PROBLEM 19 d CONTINUED

$$= \frac{1}{2(\pi)^2}\left[\cos x + x\sin x\right]_0^{2\pi}$$

$$= \frac{1}{2(\pi^2)}\left(1 + (2\pi)(0) - (1) - 0\right) = 0$$

$$b_1 = \frac{1}{\pi}\int_0^{2\pi} f(t)\sin t = \frac{1}{\pi}\int_0^{2\pi} \frac{1}{2\pi}(t)\sin t$$

$$= \frac{1}{2\pi^2}\left[\sin x - x\cos x\right]_0^{2\pi}$$

$$= \frac{1}{2\pi^2}\left(0 - 2\pi(1) - 0 - 0\right) = -\frac{1}{\pi}$$

SO, THE FIRST 2 TERMS ARE

$$f(t) = \frac{1}{3} - \frac{1}{\pi}\sin(t)$$

E-I-T HOMEWORK SOLUTIONS: COMPUTER SCIENCE

1 a) 101 + 011 = 1000 b) 101 + 110 = 1011 c) 101 + 100 = 1001

PARTS d-f: USE THE ONE'S COMPLEMENT AND THE END-AROUND CARRY. ALWAYS TAKE THE COMPLEMENT OF THE SMALLER NUMBER

d) $A_1^*(0100) = 1011$

1011 + 1100 = 10111

ADDING THE OVERFLOW BIT,

0111 + 0001 = −(1000)

THE ANSWER IS NEGATIVE BY INSPECTION

e) $A_1^*(1000) = 0111$

1110 + 0111 = 10101

0101 + 0001 = 0110

f) $A_1^*(010) = 101$

101 + 101 = 1010

010 + 001 = −(011)

THE ANSWER IS NEGATIVE BY INSPECTION

g) 111 × 11: 111, 111 → 10101

h) 100 × 11: 100, 100 → 1100

i) 1011 × 1101: 1011, 0000, 1011, 1011 → 10001111

2 a) 466 + 457 = 1145 b) 1007 + 6661 = 7670 c) 321 + 465 = 1006

d) $71 - 27 = 71 + A_7^*(27)$

$A_8^*(27) = 100 - 27 = 51$

$A_7^*(27) = A_8^* - 1 = 50$

71 + 50 = 141

ADDING THE OVERFLOW BIT,

41 + 01 = 42

e) $A_8^*(367) = 1000 - 367 = 411$

$A_7^* = 411 - 1 = 410$

1143 + 410 = 1553

553 + 001 = 554

f) TAKE THE COMPLEMENT OF THE SMALLEST NUMBER

$A_8^*(646) = 1000 - 646 = 132$

$A_7^* = 132 - 1 = 131$

677 + 131 = 1030

030 + 001 = −(031)

g) 77 × 66: 572, 572 → 6512

h) 325 × 36: 2376, 1177 → 14366

i) 3251 × 16.1: 3251, 23766, 3251 → 57023.1

3 a) BA + C = C6 b) BB + A = C5 c) DE + 10 + 1A = E8

d) FF − E = F1

e) 74 − 4A → 6 $(4+16)_{10}$ − 4 A → 6 $(20)_{10}$ − 4 $(10)_{10}$ = 2 A **2A**

f) FB − BF: E $(16+11)_{10}$ − B $(15)_{10}$ = 3 C

g) 4A × 3E: 40C, DE → 11EC

h) FE × EF: EE2, DE4 → ED22

i) 17 × 7A: E6, A1 → AF6

4 a) $6(8)^2 + 7(8)^1 + 4 = 444$

b) $1(2)^5 + 1(2)^3 + 1(2)^2 + 1 = 45$

c) $7(8)^2 + 3(8) + 4 + 2(\frac{1}{8}) + 6(\frac{1}{8})^2 + 2(\frac{1}{8})^3 = 476.34766$

d) $1(2)^3 + (1)(2)^1 + (1)(2)^0 + 1(\frac{1}{2}) + 1(\frac{1}{2})^2 = 11.75$

5 a) $\frac{75}{8} = 9$ REM 3; $\frac{9}{8} = 1$ REM 1; $\frac{1}{8} = 0$ REM 1 → $(113)_8$

b) $\frac{.375}{(\frac{1}{8})} = 3$ REM 0 → $(.3)_8$

c) $\frac{121}{8} = 15$ REM 1; $\frac{15}{8} = 1$ REM 7; $\frac{1}{8} = 0$ REM 1; $\frac{.875}{(\frac{1}{8})} = 7.0$ → 7 → $(171.7)_8$

d) 001 011 100 . 011 100 → 1 3 4 . 3 4 → $(134.34)_8$

6 a) $\frac{83}{2} = 41$ REM 1

$\frac{41}{2} = 20$ REM 1

$\frac{20}{2} = 10$ REM 0

$\frac{10}{2} = 5$ REM 0

$\frac{5}{2} = 2$ REM 1

$\frac{2}{2} = 1$ REM 0

$\frac{1}{2} = 0$ REM 1

$(1010011)_2$

b)

$\frac{100}{2} = 50$ REM 0

$\frac{50}{2} = 25$ REM 0

$\frac{25}{2} = 12$ REM 1

$\frac{12}{2} = 6$ REM 0

$\frac{6}{2} = 3$ REM 0

$\frac{3}{2} = 1$ REM 1

$\frac{1}{2} = 0$ REM 1

$\frac{.3}{\frac{1}{2}} = 0.6$ → 0

$\frac{.6}{\frac{1}{2}} = 1.2$ → 1

$\frac{.2}{(\frac{1}{2})} = 0.4$ → 0

$\frac{.4}{\frac{1}{2}} = 0.8$ → 0

$\frac{.8}{\frac{1}{2}} = 1.6$ → 1

$\frac{.6}{(\frac{1}{2})} = 1.2$ → 1

$(1100100.010011\cdots)_2$

c)

$\frac{.97}{(\frac{1}{2})} = 1.94$

$\frac{.94}{\frac{1}{2}} = 1.88$

$\frac{.88}{\frac{1}{2}} = 1.76$

$\frac{.76}{\frac{1}{2}} = 1.52$

$\frac{.52}{\frac{1}{2}} = 1.04$

$\frac{.04}{\frac{1}{2}} = 0.08$

$(.111110\cdots)_2$

d)

3 2 1 . 4 2 2

011 010 001 . 100 010 010

$(011010001.100010010)_2$

7

START
A>B?
NO
YES
B>A?
NO
YES
B>C?
NO
YES
B<C?
NO
YES
A>C?
NO
YES
C=C+A
B=B-A
C=C+B
GO TO PLACE 1
GO TO PLACE 2

8

START
14>A>10?
NO
YES
14<A<10?
NO
YES
A=A-X
EXIT

9 EVALUATE EXPRESSIONS IN PARENTHESES

$(3*4) = 12$

$(6-9) = -3$

COMBINE

$(12) + (-3) = 9$

10 a) $J<L \longrightarrow 15<9 \longrightarrow$ FALSE

$K>(M-J) \rightarrow 4>(19-15) \rightarrow$ TRUE

FALSE OR TRUE $\longrightarrow$ TRUE

b) $J>L \longrightarrow 15>9 \rightarrow$ TRUE

$K<(M-J) \longrightarrow 4<(19-15) \rightarrow$ FALSE

TRUE OR FALSE $\longrightarrow$ TRUE

11

```
      INTEGER A, B, C
      IF (A.LT.B.AND A.LT.C) GO TO 100
      IF (B.LT.C) GO TO 80
      J = C
      GO TO 110
 80   J = B
      GO TO 110
 100  J = A
 110  WRITE (6,120) J
 120  FORMAT ('1 THE SMALLEST IS", I4)
      STOP
      END
```

12

```
      REAL A, B, C
      IF (A.LE.B.OR.A.LE.C) GO TO 100
      C = C+B
      GO TO 1
 100  IF (B.LE.A.OR.B.LE.C) GO TO 200
      C = C+A
      GO TO 2
 200  IF (B.LE.A.OR.B.GT.C) CALL EXIT
      B = B-A
      GO TO 1
 1    ~~~~
 2    ~~~~
      STOP
      END
```

13

```
      INTEGER X, A
      IF (A.LE.10.OR.A.GE.14) CALL EXIT
      A = A-X
      STOP
      END
```

NOTE: THIS CHECKS FOR THE INVERSE OF WHAT IS WANTED. IF THE INVERSE IS FOUND, THE PROGRAM EXITS.

14 ASSUME EBCDIC

a) 3 IS (1111 0011)

SINCE THERE ARE 6 (AN EVEN NUMBER) OF '1' BITS, PARITY BIT = 0

PROBLEM 14 CONTINUED

b) 6 IS (11110110) PB = 0
c) J IS (11010001) PB = 0
d) T IS (11100011) PB = 1
e) X IS (1110111) PB = 0

15

ASSUME FULL USAGE OF THE 80 COLUMNS PER CARD.

$$(\text{\# CHARACTERS}) = (1000)\text{ CARDS }(80)\text{ COLUMN} = 80{,}000$$

$$(\text{\# RECORDS}) = \frac{(80{,}000)\text{ CHARACTERS}}{(1200)\text{ CHARACTERS/RECORD}} = 67$$

$$\text{\# INTERBLOCK GAPS} = 67 - 1 = 66$$

$$\text{LENGTH PER RECORD} = \frac{(1200)\text{ CHARACTERS/RECORD}}{(800)\text{ CHARACTERS/INCH}} = 1.5''$$

ASSUME INTERBLOCK GAPS OF LENGTH .6"

$$\text{TOTAL LENGTH} = 67(1.5) + 66(.6) = 140.1''$$

16

REFER TO THE PROCEDURE ON PAGE 18-19

STEP 1 DIVIDE BY 7

$$y''' + 2y'' + 6y' + \tfrac{2}{7}y = 7x$$

STEP 2 $-y''' = 2y'' + 6y' + \tfrac{2}{7}y - 7x$

STEP 3

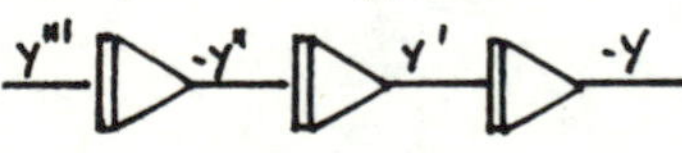

STEP 4

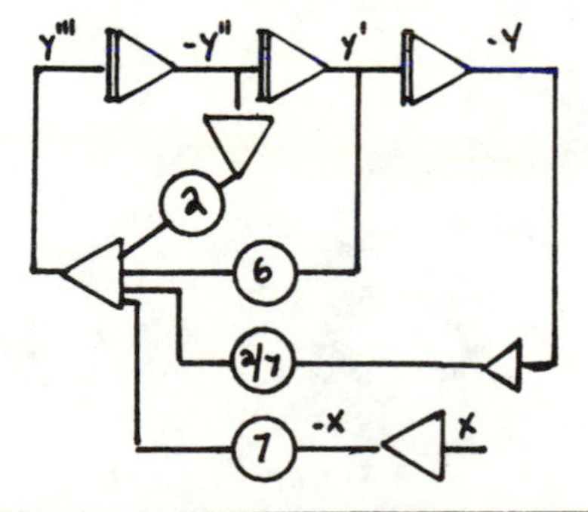

17

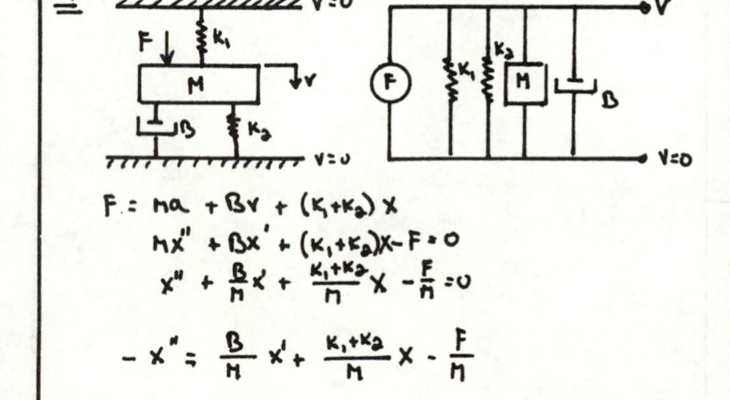

$$F = ma + Bv + (K_1+K_2)x$$

$$Mx'' + Bx' + (K_1+K_2)x - F = 0$$

$$x'' + \frac{B}{M}x' + \frac{K_1+K_2}{M}x - \frac{F}{M} = 0$$

$$-x'' = \frac{B}{M}x' + \frac{K_1+K_2}{M}x - \frac{F}{M}$$

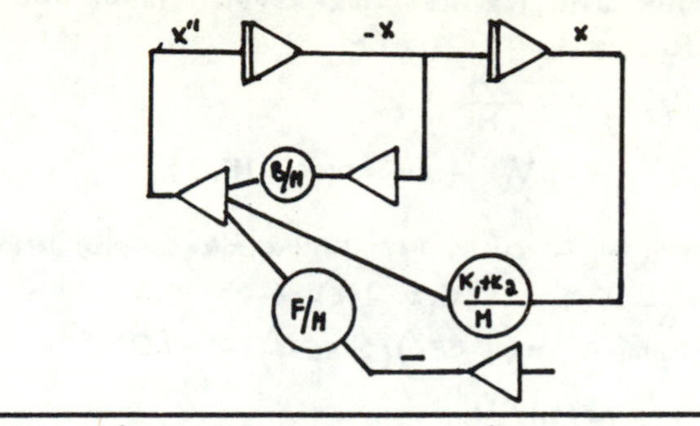

18

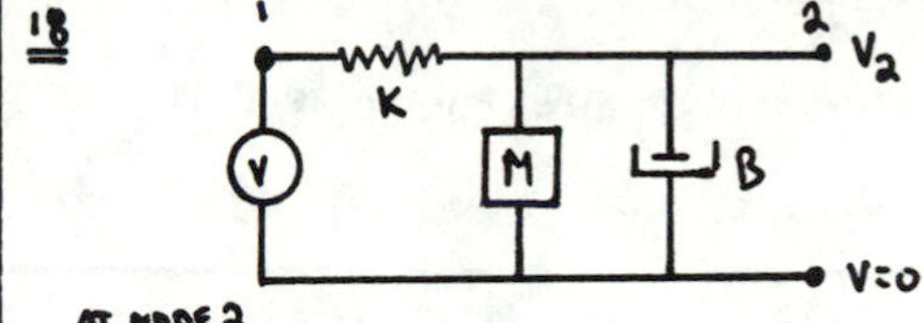

AT NODE 2

$$0 = B(v_2) + M(a_2) + K(x_2 - x)$$

$$Mx_2'' + Bx_2' + Kx_2' - Kx_1 = 0$$

$$-x_2'' = \frac{B}{M}x_2' + \frac{K}{M}x_2 - \frac{K}{M}x_1$$

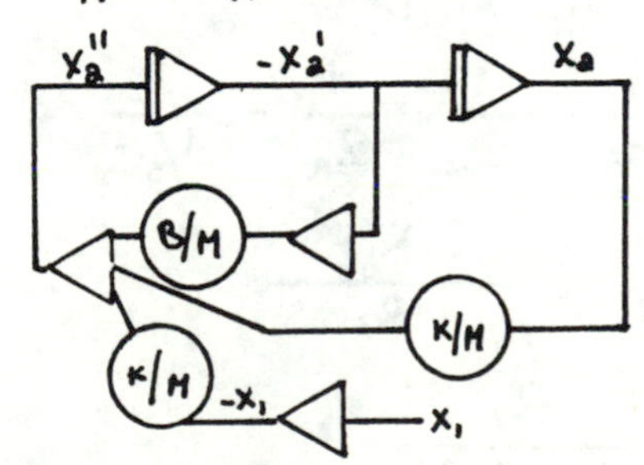

19

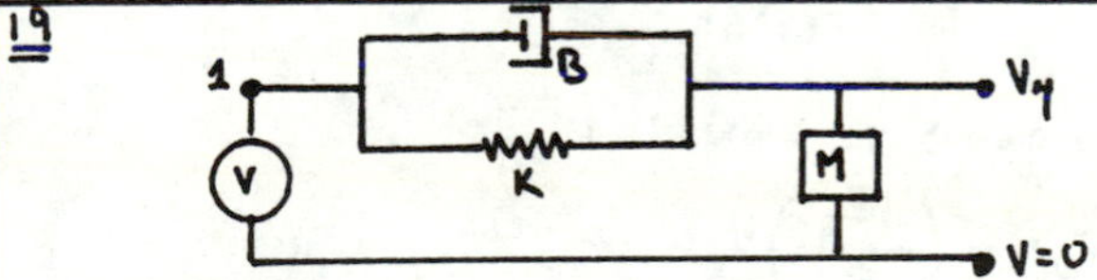

AT NODE 4

$$0 = Ma_4 + K(x_4 - x_1) + B(v_4 - v_1)$$

SO $Mx_4'' + Bx_4' + Kx_4 - Bx_1' - Kx_1 = 0$

OR $-x_4'' = \frac{B}{M}(x_4') + \frac{K}{M}x_4 - \frac{B}{M}x_1' - \frac{K}{M}x_1$

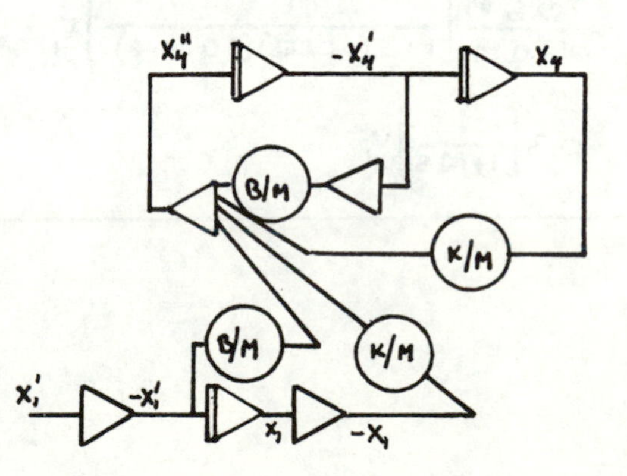

20 WORK WITH THE RESISTOR-COUPLED INPUT ONLY.

$R_f = .5\text{M} \qquad R_1 = 1\text{M}$

$R_f/R_1 = \frac{.5\text{M}}{1\text{M}} = .5$

$V_2 = -.5V_1$ {EQUATION 18.14}

NOW, WORK WITH THE CAPACITOR-COUPLED INPUT.

$R_f = .5\text{M} \qquad C_1 = 2\text{ EE-6}$

$-R_f C_1 = -(.5\text{ EE6})(2\text{ EE-6}) = -1.0$

$V_2 = (-1.0)\frac{dV_1}{dt}$

USING SUPERPOSITION, THE TOTAL OUTPUT IS

$V_2 = -.5V_1 - V_1'$

21 CONSIDER FIGURE 18.7 WHERE H IS A PARALLEL COMBINATION OF R_f AND C_f.

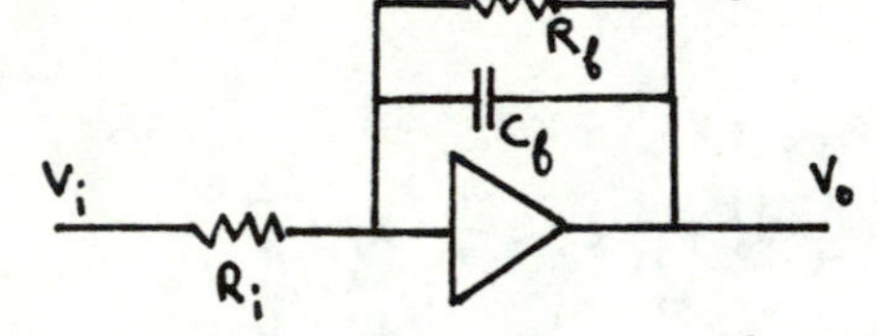

$$Y_f = \frac{1}{Z_f} = \frac{1}{Z_C} + \frac{1}{Z_R} = \frac{1}{1/sC_f} + \frac{1}{R_f}$$

$$= \frac{sC_f R_f + 1}{R_f}$$

$$Z_f = \frac{R_f}{1 + sC_f R_f}$$

ANALOGOUS TO EQN 18.14,

$$v_o = -\left(\frac{Z_f}{R_i}\right)v_i$$

$$= -\frac{R_f}{R_1}\left[\frac{1}{1 + sC_f R_f}\right]v_i$$

SUBSTITUTING VALUES,

$$v_o = -\frac{0.6\text{ M}}{2.0\text{ M}}\left[\frac{1}{1 + s(0.6\text{ EE6})(2.0\text{ EE-6})}\right]v_i$$

$$= -0.3\left[\frac{1}{1 + 1.2s}\right]v_i$$

E-I-T HOMEWORK SOLUTIONS: NUCLEONICS

1 FROM PAGE 1-38 FOR AN ELECTRON,

$M_0 = 9.11$ EE-31 KG

FROM EQN 19.42

$E_0 = M_0 C^2 = (9.11 \text{ EE-31})(3 \text{EE}8)^2$
$= 8.199$ EE-14 J

FROM PAGE 19.16

$E_0 = \dfrac{8.199 \text{ EE-14 J}}{1.60219 \text{ EE-19 J/eV}} = 5.117 \text{ EE } 5 \text{ eV}$

FROM EQN 19.78

$E_{total} = E_0 + E_k$
$= 5.117 \text{ EE5} + 90 = 5.1179 \text{ EE5}$

OR $K = \dfrac{E_{total}}{E_0} = \dfrac{5.1179}{5.117} = 1.00018$

FROM EQN 19.82

$$v = 3 \text{ EE8} \sqrt{1 - \left(\frac{1}{1.00018}\right)^2} = 5.69 \text{ EE6 M/s}$$

FROM EQN 19.34, THE WAVE VELOCITY IS

$$V_W = \frac{c^2}{v} = \frac{(3 \text{ EE8})^2}{5.69 \text{ EE6}} = 1.58 \text{ EE10 M/s}$$

FROM EQN 19.34

$$p = \frac{E_{total}}{V_W} = \frac{5.1179 \text{ EE5 eV}}{(1.58 \text{ EE10}) \frac{M}{S} (100) \frac{cM}{M}}$$
$$= 3.239 \text{ EE-7 } \frac{\text{eV-sec}}{\text{cM}}$$

ALSO FROM EQN 19.34

$$\lambda = \frac{h}{p} = \frac{4.136 \text{ EE-15 eV-sec}}{3.239 \text{ EE-7 } \frac{\text{eV-sec}}{\text{cM}}}$$
$$= 1.277 \text{ EE-8 cM}$$

2 THE EMITTED RADIATION HAS A WAVELENGTH OF

$\lambda = 6563$ EE-10 M

SINCE $c = \lambda f$,

$$f = \frac{(3 \text{ EE } 8) \text{ M/s}}{(6563 \text{ EE-10}) \text{ M}} = 4.57 \text{ EE 14 HZ}$$

FROM EQN 19.8

$\Delta E = (4.136 \text{ EE-15}) \text{ eV-sec } (4.57 \text{ EE14}) \text{ HZ}$
$= 1.89$ eV

A PHOTON AND A QUANTUM ARE IDENTICAL

3 FROM EQN 19.8

$E_2 - E_1 = hf$ A

AND $c = \lambda f$

SO $\Delta E = \dfrac{hc}{\lambda} = \dfrac{(4.136 \text{ EE-15}) \text{eV-sec} (3 \text{ EE8}) \text{ M/sec}}{(1.216 \text{ EE-10}) \text{ M}}$

$= 10.2$ eV

FROM EQN 19.14

$$\Delta E = 13.6\left(\frac{1}{(1)^2} - \frac{1}{(2)^2}\right) = 10.2 \text{eV}$$

4

$$V_W = \frac{(3 \text{ EE8})^2}{(2 \text{ EE7})} = 4.5 \text{ EE9 M/s}$$

FROM EQN 19.63

$$K = \frac{1}{\sqrt{1 - \left(\frac{2 \text{ EE7}}{3 \text{ EE8}}\right)^2}} = 1.00223$$

$$E_{total} = \frac{(9.11 \text{ EE-31}) \text{ KG} (1.00223)(3 \text{ EE8})^2 (\text{M/s})^2}{1.60219 \text{ EE-19 J/eV}}$$
$$= 5.1288 \text{ EE5 eV}$$

FROM EQN 19.34

$$p = \frac{E_{total}}{V_W} = \frac{5.1288 \text{ EE5 eV}}{(4.5 \text{ EE9}) \frac{M}{S} (100) \text{cM/M}}$$
$$= 1.1397 \text{ EE6 } \frac{\text{eV-sec}}{\text{cM}}$$

FROM EQN 19.34

$$\lambda = \frac{h}{p} = \frac{(4.136 \text{ EE-15}) \text{ eV-sec}}{(1.1397 \text{ EE6}) \frac{\text{eV-sec}}{\text{cM}}}$$
$$= 3.629 \text{ EE-9 cM}$$

5 NEUTRON

6 FROM PAGE 1-38, THE PROTON MASS IS 1.673 EE-27 KG

THE ENERGY RELEASE FROM 2 PROTONS IS

$$\frac{(2) \text{ PARTICLES } (1.673 \text{ EE-27}) \text{ KG } (3 \text{ EE8})^2 (\text{M/sec})^2}{(1.60219 \text{ EE-19}) \text{ J/eV}}$$
$$= 1.88 \text{ EE9 eV}$$

IF THE ENERGY IS EVENLY DIVIDED AMONG THE PIONS,

$$E_{PION} = \frac{1.88 \text{ EE9}}{3} = 6.271 \text{ EE8 eV}$$

7

$$\epsilon_p = \frac{h}{\epsilon_x} = \frac{(6.626 \text{ EE-34}) \text{ J-SEC}}{(.0005) \text{ M}}$$
$$= 1.325 \text{ EE-30 } \frac{\text{KG-M}}{\text{SEC}}$$

8 FROM EQN 19.37

$\epsilon_x \epsilon_p = h = 6.626$ EE-34 J-sec

BUT $\text{Joule} = \dfrac{\text{KG-M}^2}{\text{sec}^2}$

$\epsilon_p = \epsilon_{MV} = M\Delta V$
$= (9.11 \text{ EE-31}) \text{ KG} (.1) \text{ M/sec} = 9.11 \text{ EE-32 } \frac{\text{KG-M}}{\text{SEC}}$

{MORE}

PROBLEM 8 CONTINUED

$(e_x)M\,(9.11\ EE{-}32)\ \frac{KG{-}M}{SEC} = 6.626\ EE{-}34\ \frac{KG{-}M^2}{SEC}$

$e_x = 7.27\ EE{-}3\ M$

9 $\Delta M = 4.00387 - 4(1.008145) = -.02871$ AMU

FROM PAGE 19-16

$(.02871)\ AMU\ (931.481\ EE6)\ \frac{eV}{AMU} =$

$2.674\ EE7\ eV$

10 $\Delta M = 17.0045 + 1.00815 - 4.00387 - 14.0075$

$= .00128$ AMU

FROM PAGE 19-16

$(.00128)\ AMU\ (931.481\ EE6)\ \frac{eV}{AMU}$

$= 1.19\ EE6\ eV$

11 FROM EQN 19.58

$\lambda = \frac{.693}{15} = .0462\ 1/HR$

FROM EQN 19.55

$\frac{9}{48} = e^{-.0462t}$

$t = 36.23$ HRS

12 FROM EQN 19.58

$\lambda = \frac{.693}{4} = .1733\ 1/MONTH$

$p\{DECAYING\} = 1 - p\{REMAINING\}$

$= 1 - e^{-\lambda t}$

a) $\frac{M_4}{M_o} = 1 - e^{-.1733(4)} = .5$

b) $\frac{M_8}{M_o} = 1 - e^{-.1733(8)} = .75$

$.75 - .5 = .25$

c) $\frac{M_{12}}{M_o} = 1 - e^{-.1733(12)} = .875$

13 FROM EQN 19.58

$\lambda = \frac{.693}{1620} = .0004278\ 1/YR$

$1 - e^{-(.0004278)(4000)} = .819$

14 SINCE THE SAME PROPORTION OF ATOMS DECAY EACH MINUTE, IT IS APPARENT THAT THE NUMBER OF ATOMS IS HALVING EACH MINUTE.

$t_{1/2}$: 1 MINUTE

15 a) FROM EQN 19.63

$K = \frac{1}{\sqrt{1 - (.4)^2}} = 1.0911$

FROM EQN 19.42 AND EQN 19.64

$E = (1.0911)(1.673\ EE{-}27)\ KG\ (3\ EE8)^2 (M/S)^2$

$= 1.643\ EE{-}10\ J$

b) $K = \frac{1}{\sqrt{1 - (.8)^2}} = 1.667$

$E = (1.667)(9.11\ EE{-}31)(3\ EE8)^2$

$= 1.367\ EE{-}13\ J$

c) A PHOTON HAS NO REST MASS. IF THE PHOTON HAS A WAVELENGTH OF 5000 Å, THEN FROM EQN 19.35

$E_{total} = \frac{hc}{\lambda} = \frac{(4.136\ EE{-}15)\ eV\text{-}SEC\ (3\ EE8)\ M/S}{(5000\ EE{-}10)\ M}$

$= 2.48\ eV$

16 FROM EQN 19.68

$v^* = \frac{-.7c - .8c}{1 - \frac{(-.7c)(.8c)}{c^2}} = -.962\ C$

17 $\Delta E = 985\ EE6 - 938\ EE6 = 47\ EE6\ eV$

FROM PAGE 19-16

$\frac{(47\ EE6)\ eV\ (1.66053\ EE{-}27)\ \frac{KG}{AMU}}{(931.481\ EE6)\ eV/AMU}$

$= 8.379\ EE{-}29\ KG$

18 $K = \frac{140 + 35}{140} = 1.25$

FROM EQN 19.66

$t = (1.25)(1.9\ EE{-}16) = 2.375\ EE{-}16\ SEC$

19 FROM PAGE 19-16 THE PROTON REST MASS IS

$(1.007277)\ AMU\ (1.66053\ EE{-}27)\ \frac{KG}{AMU}$

$= 1.6726\ EE{-}27\ KG$

THE MASS CORRESPONDING TO THE IMPARTED ENERGY IS

$\frac{(30\ EE9)\ eV\ (1.66053\ EE{-}27)\frac{KG}{AMU}}{(931.481\ EE6)\ \frac{eV}{AMU}} = 5.348\ EE{-}26$

$K = \frac{(1.6726\ EE{-}27) + (5.348\ EE{-}26)}{1.6726\ EE{-}27} = 32.974$

FROM EQN 19.64

$M_V = 32.974\,(1.6726\ EE{-}27) = 5.5152\ EE{-}26\ KG$

FROM EQN 19.82

$V = (3\ EE10)\ \frac{CM}{SEC}\sqrt{1 - \left(\frac{1}{32.974}\right)^2} = 2.9986\ EE10\ \frac{CM}{SEC}$

{MORE}

PROBLEM 19 CONTINUED

FROM EQN 19.71

$$r = \frac{(5.5152\ EE\text{-}26)\ KG\ (1000)\ \frac{g}{KG}\ (2.9986\ EE10)\ \frac{CM}{SEC}}{(8000)\ GAUSS\ (1.602\ EE\text{-}19)\ coul\ (EE\text{-}1)\ \frac{ABcoul}{coul}}$$

$= 1.29\ EE4\ CM$

$d = 2r = 2.58\ EE4\ CM$

20 FROM EQN 19.63

$$K = \frac{1}{\sqrt{1-(.6)^2}} = 1.25$$

$\Delta M = K - 1 = .25$ (25%)

FROM EQNS 19.74 AND 19.77

$$f = \frac{(1.602\ EE\text{-}19)\ coul\ (EE\text{-}1)\ \frac{ABcoul}{coul}\ (18000)\ GAUSS}{2\pi\ (1.25)(1.673\ EE\text{-}27)\ KG\ (1000)\ \frac{g}{KG}}$$

$= 2.19\ EE7\ HZ$

21 THE ELECTRON REST MASS ENERGY IS

$(.0005486)\ AMU\ (931.481\ EE6)\ \frac{eV}{AMU}$

$= 5.11\ EE5\ eV$

$$K = \frac{5.11\ EE5 + 64}{5.11\ EE5} = 1.000125$$

FROM EQN 19.82

$$v = (3\ EE8)\sqrt{1-\left(\frac{1}{1.000125}\right)^2} = 4.74\ EE6\ \text{m/sec}$$

22 Engineer-in-Training Sample Examination

This sample Engineer-In-Training examination consists of one four-hour period. The total point value is 70 points. There are 140 questions, each worth ½ point. All 140 questions need to be answered correctly for full credit. There is no penalty for guessing or incorrect answers.

The problems have been distributed as follows:

Mathematics	1–12
Materials Science	13–18
Structure of Matter	19–23
Chemistry	24–33
Engineering Economy	34–39
Electrical Theory	40–57
Thermodynamics	58–71
Statics	72–84
Dynamics	85–97
Mechanics of Materials	98–110
Fluids	111–124
Computer Science	125–132
Systems Engineering	133–140

When permission has been given by your proctor, you should break the seal on this examination booklet and remove the answer sheet. You should immediately place your name on the answer sheet in the space indicated. You should then check that all pages are present and that each problem is legible. If any part of this booklet is missing, your proctor will issue you a completely new booklet after the old booklet has been turned in.

After you have checked your booklet thoroughly, you may begin working problems. For each problem, five answer choices lettered *A* through *E* are given from which you are to choose the one *BEST* answer. You are to record all answers on the answer sheet. No credit will be given for answers marked in the test booklet.

Use only a *Mark-Sense* or number 2 pencil. Do not use pen. Make sure that your marks are black and completely fill the bubbles. You may record only one answer per problem. If you mark more than one answer, you will not receive credit for the problem. There should be no other pencil marks on the answer sheet. If you make an error during the examination and wish to change an answer, please erase completely and carefully. Incomplete erasures may be read as intended answers.

Communicating with other examinees and smoking are not allowed during the examination. You are free to leave the room for personal reasons at any time. With your proctor's permission you may smoke outside. Do not leave the immediate area. Do not talk to other examinees outside the room. You may not go to your car or office for additional reference material.

This examination is open book. You may use any quantity of permanently bound, spiral-bound, or three-ring bound references. However, you may not use loose paper, scratch pads, blank tablets, or booklets. You may use the margins and any other part of this booklet for computations. A sheet of graph paper is provided in this booklet.

Your score will be based on the number of questions you answer correctly. Use your time effectively and mark the best answer you can find for every question. Do not waste time on questions that are too difficult for you.

You will be told when you have approximately 10 minutes remaining.

PLEASE WAIT FOR PERMISSION TO BEGIN.

You may now ask questions.

1 Ⓐ Ⓑ Ⓒ Ⓓ Ⓔ	26 Ⓐ Ⓑ Ⓒ Ⓓ Ⓔ	51 Ⓐ Ⓑ Ⓒ Ⓓ Ⓔ	76 Ⓐ Ⓑ Ⓒ Ⓓ Ⓔ	101 Ⓐ Ⓑ Ⓒ Ⓓ Ⓔ	126 Ⓐ Ⓑ Ⓒ Ⓓ Ⓔ
2 Ⓐ Ⓑ Ⓒ Ⓓ Ⓔ	27 Ⓐ Ⓑ Ⓒ Ⓓ Ⓔ	52 Ⓐ Ⓑ Ⓒ Ⓓ Ⓔ	77 Ⓐ Ⓑ Ⓒ Ⓓ Ⓔ	102 Ⓐ Ⓑ Ⓒ Ⓓ Ⓔ	127 Ⓐ Ⓑ Ⓒ Ⓓ Ⓔ
3 Ⓐ Ⓑ Ⓒ Ⓓ Ⓔ	28 Ⓐ Ⓑ Ⓒ Ⓓ Ⓔ	53 Ⓐ Ⓑ Ⓒ Ⓓ Ⓔ	78 Ⓐ Ⓑ Ⓒ Ⓓ Ⓔ	103 Ⓐ Ⓑ Ⓒ Ⓓ Ⓔ	128 Ⓐ Ⓑ Ⓒ Ⓓ Ⓔ
4 Ⓐ Ⓑ Ⓒ Ⓓ Ⓔ	29 Ⓐ Ⓑ Ⓒ Ⓓ Ⓔ	54 Ⓐ Ⓑ Ⓒ Ⓓ Ⓔ	79 Ⓐ Ⓑ Ⓒ Ⓓ Ⓔ	104 Ⓐ Ⓑ Ⓒ Ⓓ Ⓔ	129 Ⓐ Ⓑ Ⓒ Ⓓ Ⓔ
5 Ⓐ Ⓑ Ⓒ Ⓓ Ⓔ	30 Ⓐ Ⓑ Ⓒ Ⓓ Ⓔ	55 Ⓐ Ⓑ Ⓒ Ⓓ Ⓔ	80 Ⓐ Ⓑ Ⓒ Ⓓ Ⓔ	105 Ⓐ Ⓑ Ⓒ Ⓓ Ⓔ	130 Ⓐ Ⓑ Ⓒ Ⓓ Ⓔ
6 Ⓐ Ⓑ Ⓒ Ⓓ Ⓔ	31 Ⓐ Ⓑ Ⓒ Ⓓ Ⓔ	56 Ⓐ Ⓑ Ⓒ Ⓓ Ⓔ	81 Ⓐ Ⓑ Ⓒ Ⓓ Ⓔ	106 Ⓐ Ⓑ Ⓒ Ⓓ Ⓔ	131 Ⓐ Ⓑ Ⓒ Ⓓ Ⓔ
7 Ⓐ Ⓑ Ⓒ Ⓓ Ⓔ	32 Ⓐ Ⓑ Ⓒ Ⓓ Ⓔ	57 Ⓐ Ⓑ Ⓒ Ⓓ Ⓔ	82 Ⓐ Ⓑ Ⓒ Ⓓ Ⓔ	107 Ⓐ Ⓑ Ⓒ Ⓓ Ⓔ	132 Ⓐ Ⓑ Ⓒ Ⓓ Ⓔ
8 Ⓐ Ⓑ Ⓒ Ⓓ Ⓔ	33 Ⓐ Ⓑ Ⓒ Ⓓ Ⓔ	58 Ⓐ Ⓑ Ⓒ Ⓓ Ⓔ	83 Ⓐ Ⓑ Ⓒ Ⓓ Ⓔ	108 Ⓐ Ⓑ Ⓒ Ⓓ Ⓔ	133 Ⓐ Ⓑ Ⓒ Ⓓ Ⓔ
9 Ⓐ Ⓑ Ⓒ Ⓓ Ⓔ	34 Ⓐ Ⓑ Ⓒ Ⓓ Ⓔ	59 Ⓐ Ⓑ Ⓒ Ⓓ Ⓔ	84 Ⓐ Ⓑ Ⓒ Ⓓ Ⓔ	109 Ⓐ Ⓑ Ⓒ Ⓓ Ⓔ	134 Ⓐ Ⓑ Ⓒ Ⓓ Ⓔ
10 Ⓐ Ⓑ Ⓒ Ⓓ Ⓔ	35 Ⓐ Ⓑ Ⓒ Ⓓ Ⓔ	60 Ⓐ Ⓑ Ⓒ Ⓓ Ⓔ	85 Ⓐ Ⓑ Ⓒ Ⓓ Ⓔ	110 Ⓐ Ⓑ Ⓒ Ⓓ Ⓔ	135 Ⓐ Ⓑ Ⓒ Ⓓ Ⓔ
11 Ⓐ Ⓑ Ⓒ Ⓓ Ⓔ	36 Ⓐ Ⓑ Ⓒ Ⓓ Ⓔ	61 Ⓐ Ⓑ Ⓒ Ⓓ Ⓔ	86 Ⓐ Ⓑ Ⓒ Ⓓ Ⓔ	111 Ⓐ Ⓑ Ⓒ Ⓓ Ⓔ	136 Ⓐ Ⓑ Ⓒ Ⓓ Ⓔ
12 Ⓐ Ⓑ Ⓒ Ⓓ Ⓔ	37 Ⓐ Ⓑ Ⓒ Ⓓ Ⓔ	62 Ⓐ Ⓑ Ⓒ Ⓓ Ⓔ	87 Ⓐ Ⓑ Ⓒ Ⓓ Ⓔ	112 Ⓐ Ⓑ Ⓒ Ⓓ Ⓔ	137 Ⓐ Ⓑ Ⓒ Ⓓ Ⓔ
13 Ⓐ Ⓑ Ⓒ Ⓓ Ⓔ	38 Ⓐ Ⓑ Ⓒ Ⓓ Ⓔ	63 Ⓐ Ⓑ Ⓒ Ⓓ Ⓔ	88 Ⓐ Ⓑ Ⓒ Ⓓ Ⓔ	113 Ⓐ Ⓑ Ⓒ Ⓓ Ⓔ	138 Ⓐ Ⓑ Ⓒ Ⓓ Ⓔ
14 Ⓐ Ⓑ Ⓒ Ⓓ Ⓔ	39 Ⓐ Ⓑ Ⓒ Ⓓ Ⓔ	64 Ⓐ Ⓑ Ⓒ Ⓓ Ⓔ	89 Ⓐ Ⓑ Ⓒ Ⓓ Ⓔ	114 Ⓐ Ⓑ Ⓒ Ⓓ Ⓔ	139 Ⓐ Ⓑ Ⓒ Ⓓ Ⓔ
15 Ⓐ Ⓑ Ⓒ Ⓓ Ⓔ	40 Ⓐ Ⓑ Ⓒ Ⓓ Ⓔ	65 Ⓐ Ⓑ Ⓒ Ⓓ Ⓔ	90 Ⓐ Ⓑ Ⓒ Ⓓ Ⓔ	115 Ⓐ Ⓑ Ⓒ Ⓓ Ⓔ	140 Ⓐ Ⓑ Ⓒ Ⓓ Ⓔ
16 Ⓐ Ⓑ Ⓒ Ⓓ Ⓔ	41 Ⓐ Ⓑ Ⓒ Ⓓ Ⓔ	66 Ⓐ Ⓑ Ⓒ Ⓓ Ⓔ	91 Ⓐ Ⓑ Ⓒ Ⓓ Ⓔ	116 Ⓐ Ⓑ Ⓒ Ⓓ Ⓔ	
17 Ⓐ Ⓑ Ⓒ Ⓓ Ⓔ	42 Ⓐ Ⓑ Ⓒ Ⓓ Ⓔ	67 Ⓐ Ⓑ Ⓒ Ⓓ Ⓔ	92 Ⓐ Ⓑ Ⓒ Ⓓ Ⓔ	117 Ⓐ Ⓑ Ⓒ Ⓓ Ⓔ	
18 Ⓐ Ⓑ Ⓒ Ⓓ Ⓔ	43 Ⓐ Ⓑ Ⓒ Ⓓ Ⓔ	68 Ⓐ Ⓑ Ⓒ Ⓓ Ⓔ	93 Ⓐ Ⓑ Ⓒ Ⓓ Ⓔ	118 Ⓐ Ⓑ Ⓒ Ⓓ Ⓔ	
19 Ⓐ Ⓑ Ⓒ Ⓓ Ⓔ	44 Ⓐ Ⓑ Ⓒ Ⓓ Ⓔ	69 Ⓐ Ⓑ Ⓒ Ⓓ Ⓔ	94 Ⓐ Ⓑ Ⓒ Ⓓ Ⓔ	119 Ⓐ Ⓑ Ⓒ Ⓓ Ⓔ	
20 Ⓐ Ⓑ Ⓒ Ⓓ Ⓔ	45 Ⓐ Ⓑ Ⓒ Ⓓ Ⓔ	70 Ⓐ Ⓑ Ⓒ Ⓓ Ⓔ	95 Ⓐ Ⓑ Ⓒ Ⓓ Ⓔ	120 Ⓐ Ⓑ Ⓒ Ⓓ Ⓔ	
21 Ⓐ Ⓑ Ⓒ Ⓓ Ⓔ	46 Ⓐ Ⓑ Ⓒ Ⓓ Ⓔ	71 Ⓐ Ⓑ Ⓒ Ⓓ Ⓔ	96 Ⓐ Ⓑ Ⓒ Ⓓ Ⓔ	121 Ⓐ Ⓑ Ⓒ Ⓓ Ⓔ	
22 Ⓐ Ⓑ Ⓒ Ⓓ Ⓔ	47 Ⓐ Ⓑ Ⓒ Ⓓ Ⓔ	72 Ⓐ Ⓑ Ⓒ Ⓓ Ⓔ	97 Ⓐ Ⓑ Ⓒ Ⓓ Ⓔ	122 Ⓐ Ⓑ Ⓒ Ⓓ Ⓔ	
23 Ⓐ Ⓑ Ⓒ Ⓓ Ⓔ	48 Ⓐ Ⓑ Ⓒ Ⓓ Ⓔ	73 Ⓐ Ⓑ Ⓒ Ⓓ Ⓔ	98 Ⓐ Ⓑ Ⓒ Ⓓ Ⓔ	123 Ⓐ Ⓑ Ⓒ Ⓓ Ⓔ	
24 Ⓐ Ⓑ Ⓒ Ⓓ Ⓔ	49 Ⓐ Ⓑ Ⓒ Ⓓ Ⓔ	74 Ⓐ Ⓑ Ⓒ Ⓓ Ⓔ	99 Ⓐ Ⓑ Ⓒ Ⓓ Ⓔ	124 Ⓐ Ⓑ Ⓒ Ⓓ Ⓔ	
25 Ⓐ Ⓑ Ⓒ Ⓓ Ⓔ	50 Ⓐ Ⓑ Ⓒ Ⓓ Ⓔ	75 Ⓐ Ⓑ Ⓒ Ⓓ Ⓔ	100 Ⓐ Ⓑ Ⓒ Ⓓ Ⓔ	125 Ⓐ Ⓑ Ⓒ Ⓓ Ⓔ	

Your name: ______________________________ No. correct __________

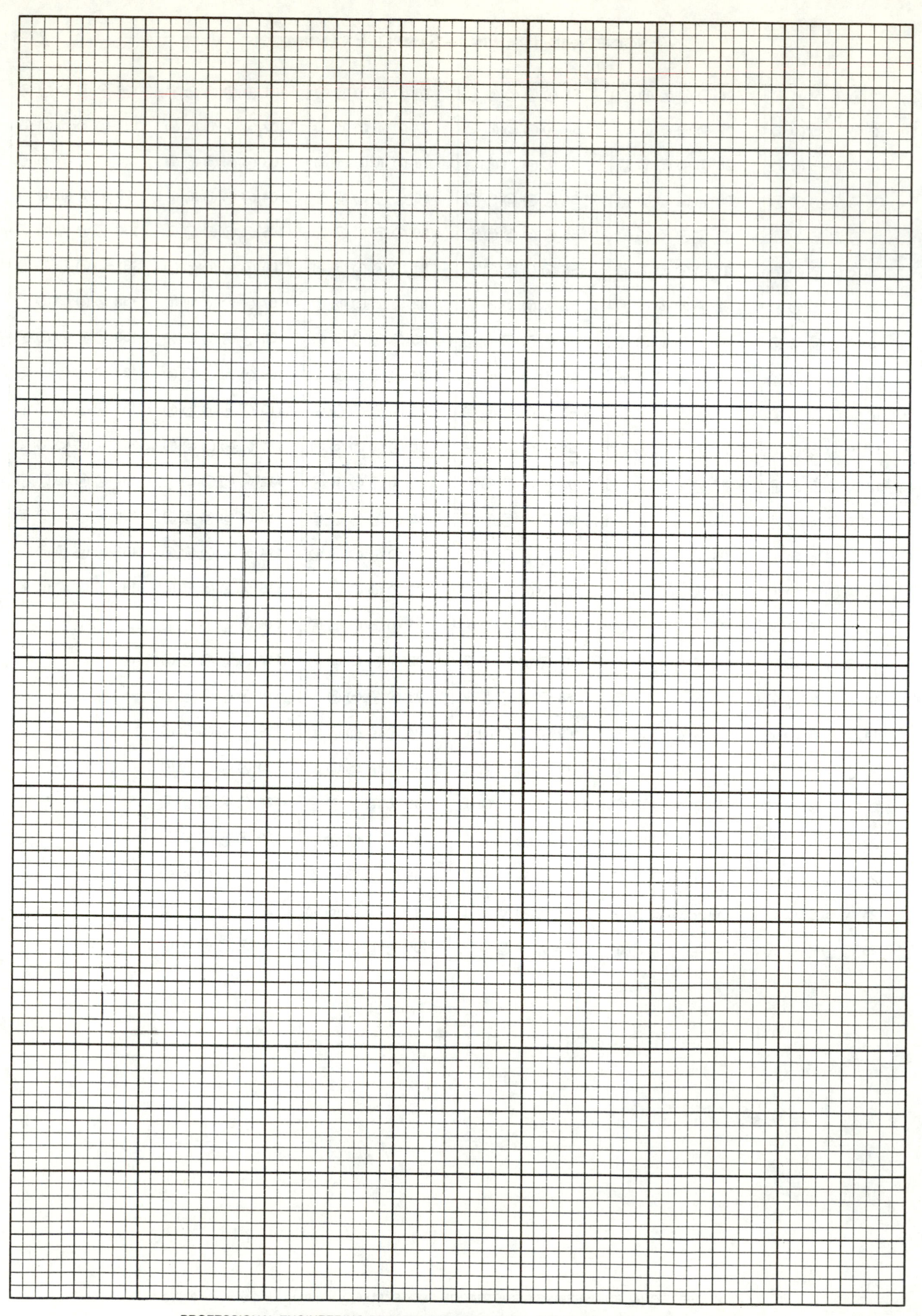

MATHEMATICS

1. Five fair coins are each flipped once. What is the probability that at least two coins will show heads?

(A) .187 (B) .80 (C) .156
(D) .844 (E) .813

2. What is the determinant of the following (2 x 2) matrix?

$$\begin{bmatrix} 1 & 4 \\ 3 & 2 \end{bmatrix}$$

(A) 10 (B) 14 (C) -5
(D) -10 (E) 9

3. What is the radius of a circle with the equation given below?

$$x^2 - 6x + y^2 - 4y - 12 = 0$$

(A) 4 (B) 5 (C) 7
(D) 6 (E) 3.46

4. What is the value of the following limit?

$$\lim_{x \to 3} \frac{(x^2 - 9)}{(x-3)}$$

(A) 1 (B) 9 (C) -6
(D) 6 (E) ∞

5. Given the vector **V** = i + 2j + k, what is the angle between **V** and the x axis?

(A) 22° (B) 24° (C) 1.15°
(D) 80° (E) 66°

6. Evaluate the following definite integral:

$$\int_2^{\infty} (1/x^2)dx$$

(A) .125 (B) 8 (C) 2
(D) -.5 (E) .5

7. What is the standard deviation of 1, 4, and 7?

(A) 2.45 (B) 6 (C) 32
(D) 5.66 (E) 3.0

8. Which is a true statement about the two vectors?

$$V_1 = i + 2j + k$$
$$V_2 = i + 3j - 7k$$

(A) the vectors coincide
(B) both vectors pass through (0, −1, 6)
(C) the vectors are parallel
(D) the vectors are orthogonal
(E) the angle between the vectors is 17.4°

9. What is the area bounded by $y = 0$, $y = e^x$, $x = 0$, and $x = 1$?

(A) 2.72 (B) .86 (C) 3.44
(D) 1.72 (E) 1.36

10. A function of x is given below. Which coordinates identify a relative maximum or minimum?

$$y = \frac{1}{4}x^4 - 1.5x^2 + 2x + 5$$

(A) (-2, -1) (B) (-2, -2) (C) (+2, -2)
(D) (-1, -1.75) (E) (+1, +5.75)

11. The slope of a line is ½. The slope of a second line is (-2/3). Both lines intercept at the point (3, 1). What is the acute angle between the lines?

(A) 50.2° (B) 60.3° (C) 119.7°
(D) 26.6° (E) 33.7°

12. A function is given below. What is x for y to be maximum?

$$y^2 + y + x^2 - 2x = 5$$

(A) 1 (B) -1 (C) ½
(D) 5 (E) 2.23

MATERIALS SCIENCE

13. What are the indexes of the plane shown below?

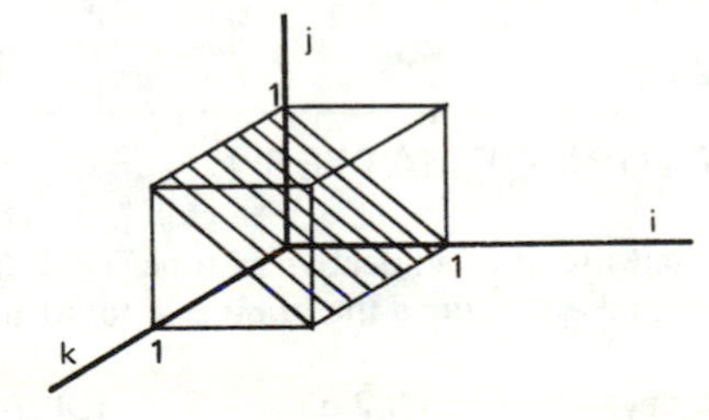

(A) [101] (B) (101) (C) (1∞1)
(D) (110) (E) (11∞)

14. How many atoms are in a hexagonal, close-packed cell?

(A) 1 (B) 2 (C) 3
(D) 4 (E) 6

15. Which type of hardening will work to some extent in all metals?

(A) work hardening
(B) annealing
(C) martempering
(D) austenitizing
(E) shot-peening

16. Which of the following is not a face-centered cubic structure?

(A) copper
(B) graphite
(C) silver
(D) nickel
(E) platinum

17. Sound travels fastest in

(A) steel at 70°F
(B) air at 0°F and one atmosphere
(C) air at 70°F and one atmosphere
(D) air at 70°F and 0 psia
(E) hydrogen at 0°F and 1 atmosphere

18. What is the distance between atoms A and B in the face-centered cubic cell shown? Assume atoms pack together as hard spheres.

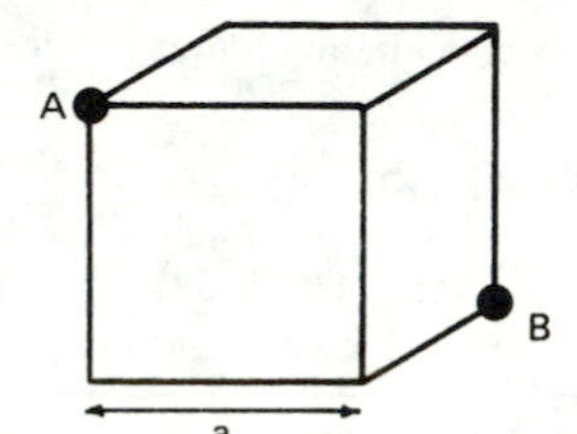

(A) $2\sqrt{6}$ a
(B) $2\sqrt{3}$ a
(C) $\sqrt{3}$ a
(D) 4a
(E) $2\sqrt{2}$ a

STRUCTURE OF MATTER

19. The half-life of a radioactive isotope is 4.3 days. How long will it take to reduce the original amount to 1%?

(A) 28.6 days (B) 3.2 days (C) .74 days
(D) 15.7 days (E) 39.7 days

20. An orbital

(A) may have 2, 8, 18, or 32 electrons.
(B) may have 2 electrons with the same spin direction.
(C) cannot exist below the critical pressure.
(D) may be photographed with an electron microscope.
(E) may be unoccupied.

21. Why are fission reactors in wider-spread use than fusion power sources?

(A) Fusion sources create radioactive wastes that cannot yet be adequately disposed of.
(B) Fission is more productive than fusion per pound of fuel.
(C) Fusion is suitable only for small-scale plants.
(D) There is insufficient fuel for fusion sources.
(E) Fusion reactions cannot yet be adequately contained.

22. The mass of a photon at .8 times the speed of light (.8c) is most nearly

(A) 2.2 amu (B) 1.7 amu (C) .8 amu
(D) .001 amu (E) 0

23. The radius of a hypothetical electron orbit is known to be .75 Å. What is the de Broglie wavelength of the electron if 4 complete cycles constitute a stable pattern around the nucleus?

(A) 9.42 Å (B) 4.71 Å (C) 2.36 Å
(D) 1.18 Å (E) .188 Å

CHEMISTRY

24. How many isomers of phthalic acid exist?

(A) 1 (B) 2 (C) 3
(D) 4 (E) 5

25. 10 grams of table salt are dissolved in 2 liters of solution. What is the molarity?

(A) .085 (B) .171 (C) .342
(D) 29.3 (E) 5

26. What is the formula of a compound with the following gravimetric analysis?

oxygen	13.7%
carbon	20.5%
hydrogen	5.1%
chlorine	60.7%

(A) CH_3OCl (B) C_2H_6OCl (C) $C_2H_6OCl_2$
(D) CH_5OCl (E) CH_6O_2Cl

27. What volume of an H_2SO_4 solution (normality = 36) is required to prepare 15 ml of H_2SO_4 with a normality of 5?

(A) 2.1 ml (B) 12 ml (C) 108 ml
(D) 15 ml (E) 5 ml

28. What is the normality of 18.6 ml of H_2SO_4 which neutralizes 30 ml of a 1.5 N KOH solution?

(A) 7.4 (B) .96 (C) .096
(D) 2.4 (E) 4.9

29. What is oxidized and what is reduced in the reaction shown below?

$$Zn + H_2SO_4 \rightarrow ZnSO_4 + H_2$$

(A) Zinc is oxidized only.
(B) Zinc is reduced only.
(C) Zinc is oxidized and hydrogen is reduced.
(D) Zinc is reduced and hydrogen is oxidized.
(E) Sulfate is oxidized and reduced.

30. What is the oxidation number of nitrogen in HNO_3?

(A) -1 (B) +1 (C) +2
(D) -2 (E) +5

31. How many moles of NaOH will be neutralized by one mole of H_3PO_4?

(A) 1/3 (B) 1 (C) 2
(D) 3 (E) 6

32. An anode is

(A) a pentode with a grounded plate.
(B) the end-point of a directed network.
(C) an electrode at which oxidation occurs.
(D) an electrode at which reduction occurs.
(E) the electrode to which the cation would be attracted during an electrolytic reaction.

33. Beryllium, magnesium, and calcium all belong to which elemental grouping?

(A) noble elements
(B) halogens
(C) lanthonons
(D) alkalai metals
(E) alkaline earth metals

ENGINEERING ECONOMY

34. A bank offers 1.2% compounded monthly. What is the effective annual rate?

(A) 14.4% (B) 15.4% (C) 115%
(D) 8.9% (E) 7.9%

35. A bank charges 12% simple interest on a $300 loan. How much will be repaid if the loan is paid back in one lump sum after 3 years?

(A) $108 (B) $408 (C) $421
(D) $415 (E) $436

36. A $10,000 loan is taken out. It is to be repaid at the rate of $200 per month, with an annual effective interest rate of 19.56% charged against the unpaid balance. What principal remains to be paid off after the third payment?

(A) $9511 (B) $9400 (C) $9763
(D) $9899 (E) $9848

37. Which of the following methods is most suited to evaluating and comparing alternatives with different (finite) lives?

(A) present worth method
(B) MAPI method
(C) uniform annual cost method
(D) rate of return method
(E) urgency rating method

38. A company purchases an asset for $10,000 which it plans to keep for 20 years. If the salvage value is zero at the end of the 20th year, what is the depreciation in the 3rd year? (Use sum-of-the-years' digits.)

(A) $952 (B) $905 (C) $857
(D) $500 (E) $1500

39. $1000 invested now is left for eight years, at which time the principal is withdrawn. The interest that has accrued is left for another 8 years. If the effective annual interest rate is 5%, what will be the withdrawal at the end of the 16th year?

(A) $1000 (B) $500 (C) $706
(D) $477 (E) $1477

ELECTRICAL THEORY

40. Six coulombs of charge pass through a wire in 2 seconds. What is the average current flowing?

(A) 3 amps (B) 6 amps (C) 12 amps
(D) 1.6 amps (e) 4.8 amps

41. A transformer has 200 primary turns and 20 secondary turns. What is the secondary voltage if the primary voltage is 120 volts?

(A) 12 volts (B) 1200 volts (C) 1.2 volts
(D) 12,000 volts (E) 120 volts

42. If .3 amps flow in the secondary and 30 amps flow in the primary of a perfectly-matched, ideal transformer, what is the primary:secondary turns ratio?

(A) 100:1 (B) 1:100 (C) 1:10
(D) 10:1 (E) 1:10,000

43. Five watts enter a perfectly-matched, ideal transformer with a primary/secondary turns ratio of 15:1. If the input impedance is 2000 ohms, what is the load?

(A) 79 ohms (B) 6243 ohms (C) 133 ohms
(D) 30,000 ohms (E) 8.9 ohms

44. What is the total energy transferred when the capacitor discharges?

R
V^+

(A) V^2R (B) ½CV (C) CV
(D) ½CV^2 (E) CV^2

45. What will be the phase angle difference between the current and the voltage. Take the voltage as the reference.

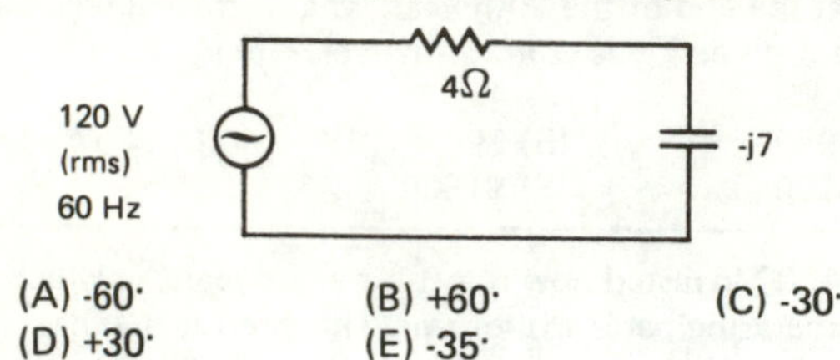

(A) -60°
(B) +60°
(C) -30°
(D) +30°
(E) -35°

46. Two 10 ohm resistances are connected in parallel. This combination is connected in series with a capacitor of 5 farads. The circuit is connected across a sinusoidal source with a frequency of 0 and a maximum voltage of 50 volts. What is the maximum current?

(A) 5 amps
(B) 0 amps
(C) 7.07 amps
(D) 1 amp
(E) none of the above

47. A circuit draws 5000 kva with a power factor of .72. What size capacitor is required to increase the power factor to .86? The 60 hz line voltage is 220 (rms).

(A) 13.95 F
(B) 157 F
(C) .0002 F
(D) .042 F
(E) .0729 F

48. What is the alpha parameter (α) in an NPN transistor amplifier?

(A) collector/emitter current ratio for common base operation
(B) emitter/collector current ratio for common base operation
(C) collector/base current ratio for common base operation
(D) collector/emitter current ratio for common emitter operation
(E) collector/base current ratio for common emitter operation

49. The voltage appearing across the 4 ohm resistor is closest to

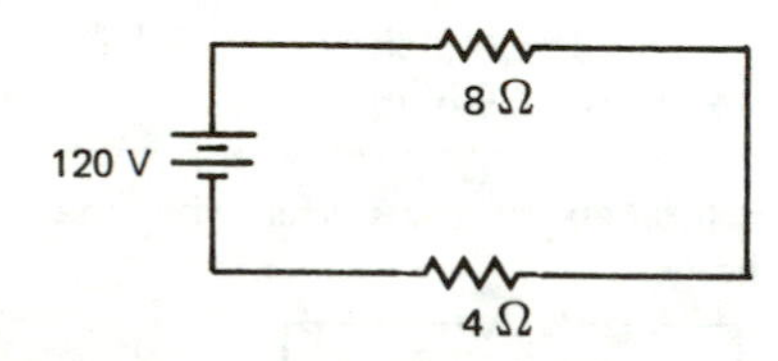

(A) 40 volts
(B) 80 volts
(C) 26.7 volts
(D) 60 volts
(E) 120 volts

50. At 60 hz, a series LCR circuit consists of X_L = 20 ohms, X_C = 14 ohms, and R = 10 ohms. What is the impedance at resonant frequency?

(A) 20 ohms
(B) 14 ohms
(C) 10 ohms
(D) 24.4 ohms
(E) 26.4 ohms

51. The correct answer to this question is given by the following logic diagram. (There is no choice 'E'.)

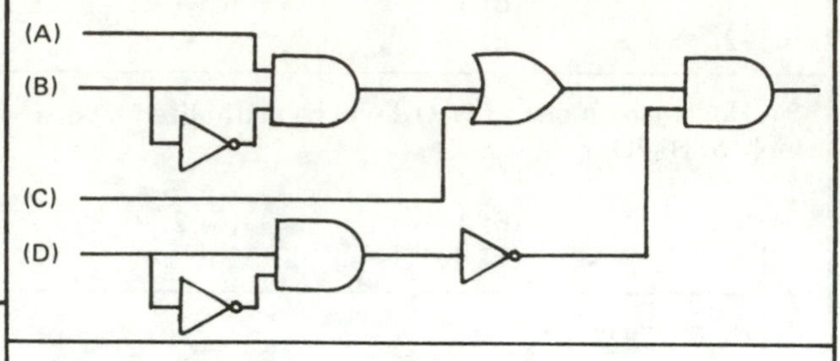

52. A square wave is put into a class B amplifier. What will be the output?

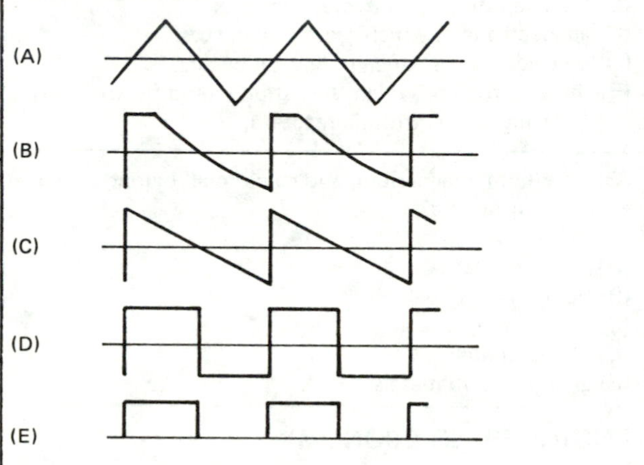

53. What is the phasor voltage drop across the source?

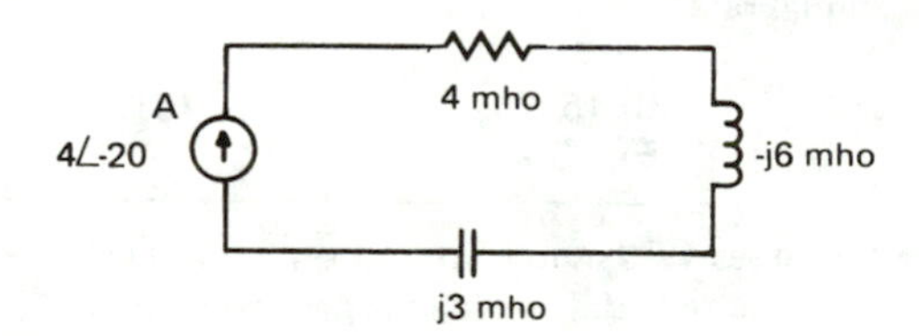

(A) .8∠16.8°
(B) 1.2∠-53.7°
(C) 5∠36.8°
(D) 6∠-20°
(E) 6∠20°

54. What is the overall gain of the cascaded system shown below?

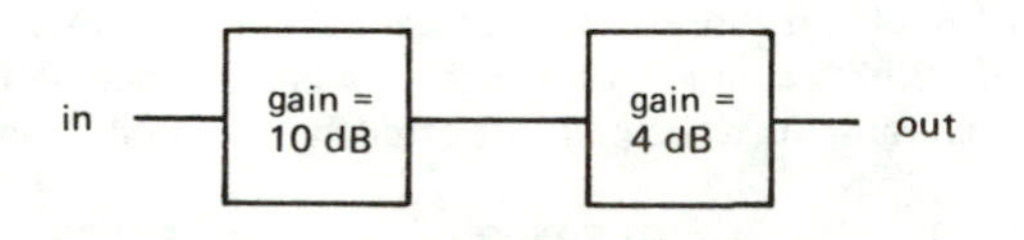

(A) 1.6 dB
(B) 16 dB
(C) 11.5 dB
(D) 14 dB
(E) 40 dB

55. The addition of gallium to pure germanium is an example of

(A) ionic bonding
(B) n-type doping
(C) p-type doping
(D) oxidation
(E) biasing

56. What is the current in the 6 ohm resistor?

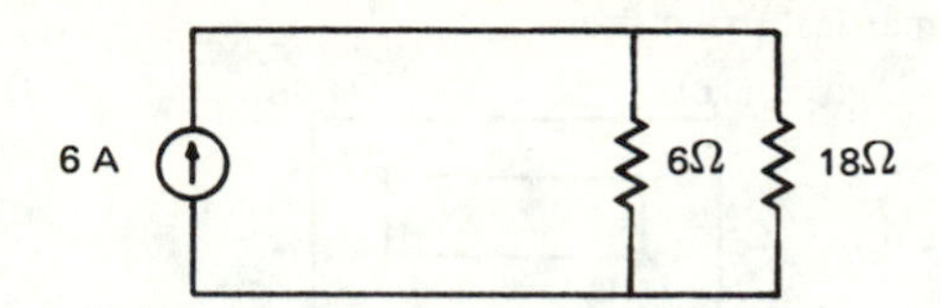

(A) 1.5 amps (B) 0 amps (C) 4 amps
(D) 4.5 amps (E) 2 amps

57. What is the Thevenin equivalent for the circuit shown below?

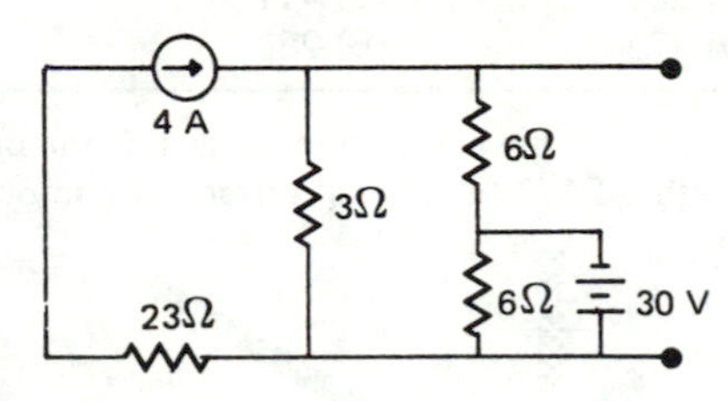

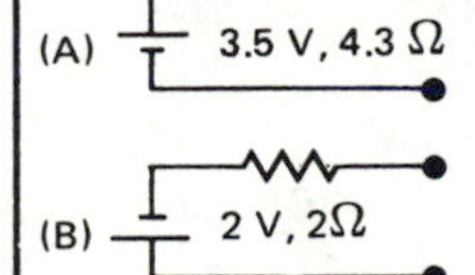

(C) 2.3 V, .5Ω

(D) 30 V, 6Ω

(E) 2 V, 3Ω

THERMODYNAMICS

58. What are the changes in internal energy for reversible adiabatic and isothermal processes respectively?

(A) $C_p\Delta T$ and 0
(B) 0 and $C_v\Delta T$
(C) $C_v\Delta T$ and $C_p\Delta T$
(D) $C_v\Delta T$ and 0
(E) 0 and 0

59. A steam engine operates between 150°C and 550°C. What is its theoretical maximum thermal efficiency?

(A) 73% (B) 49% (C) 266%
(D) 95% (E) 37%

60. What is the coefficient of performance of a Carnot refrigeration cycle operating between -10°F and -190°F?

(A) 2.5 (B) 1.5 (C) .5
(D) 1.06 (E) -.5

61. The equation $TdS = pdV + dU$ is valid

(A) only in constant temperature processes
(B) only in constant pressure processes
(C) only in reversible processes
(D) only in constant volume processes
(E) always

62. Which of the following statements is true for a perfect gas flowing through an insulated valve?

(A) Enthalpy is essentially unchanged.
(B) Entropy decreases.
(C) Temperature increases greatly.
(D) Flow is isentropic.
(E) Pressure increases.

63. If an initial volume of an ideal gas is compressed to one-half of its original volume and to twice its original temperature, the pressure

(A) remains constant
(B) doubles
(C) quadruples
(D) halves
(E) triples

64. If an initial volume of saturated steam is expanded isothermally to twice the initial volume, the pressure

(A) remains the same
(B) decreases
(C) increases
(D) halves
(E) doubles

65. An adiabatic process

(A) allows heat transfer into the system but not out of the system.
(B) allows heat transfer out of the system but not into the system.
(C) may be reversible.
(D) is one in which enthalpy remains unchanged.
(E) is one in which the equation $W = Q$ is valid.

66. What work is done in compressing 1 lbm of air at STP to one-half of its original volume in an isothermal process?

(A) -1690 BTU (B) +1690 BTU (C) -59 BTU
(D) +59 BTU (E) none of the above

67. Regardless of the process, the change in enthalpy for an ideal gas is

(A) nR^*T/V (B) 0 (C) nC_vdT
(D) nC_pdT (E) Q

68. Which of the statements is true about entropy?

(A) Entropy always decreases.
(B) Entropy increases up to the critical temperature, then it decreases.
(C) Theoretically may be zero at a low enough temperature.
(D) Doesn't change in a throttling process.
(E) None of the above statements are true.

69. A gas at one atmosphere originally occupies a volume of 20 cubic feet. The gas is polytropically compressed with n = 1.15 until the pressure is 48.51 psia. What work is required for this compression?

(A) 300 ft-lbf (B) 300 BTU (C) 48,000 ft-lbf
(D) 115 ft-lbf (E) 45 ft-lbf

70. A 48 square foot wall has an average thermal conductivity of 1.2 BTU-ft/ft^2-°F-hr. If the temperature difference is 18°F between the two sides and the thickness of .7 feet, what is the total heat loss in 5 hours?

(A) 1480 BTU (B) 7400 BTU (C) 5180 BTU
(D) 38,500 BTU (E) none of the above

71. Calculate the Joule-Thomson (Joule-Kelvin) coefficient for superheated steam at 240 psia and 450°F that is throttled to 60 psia.

(A) .28 °F/psi (B) 3.6 °F/psi (C) 0 °F/psi
(D) .073 °F/psi (E) none of the above

STATICS

72. Find the y component of the centroid of the object shown below.

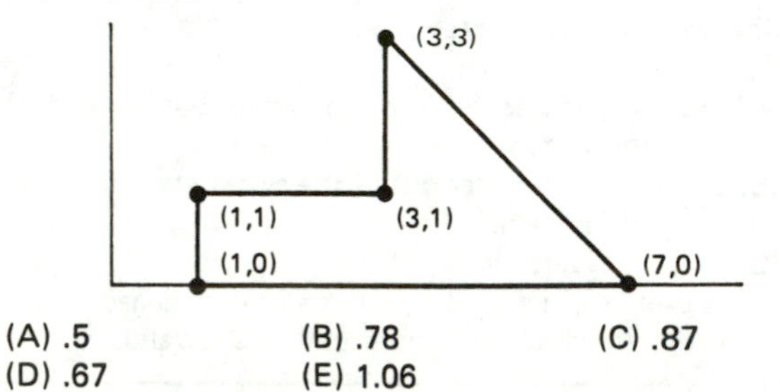

(A) .5 (B) .78 (C) .87
(D) .67 (E) 1.06

73. What is the maximum value of x so that F can be applied without tipping the block? (μ = .3.)

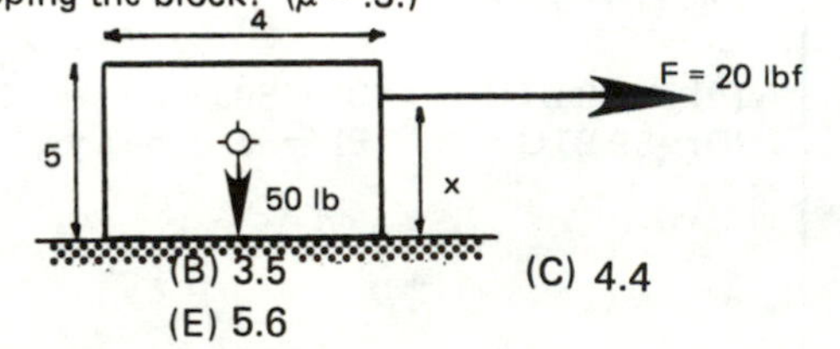

(A) 5.0 (B) 3.5 (C) 4.4
(D) 1.5 (E) 5.6

74. Find the moment of inertia about the x axis of the symmetrical shape shown.

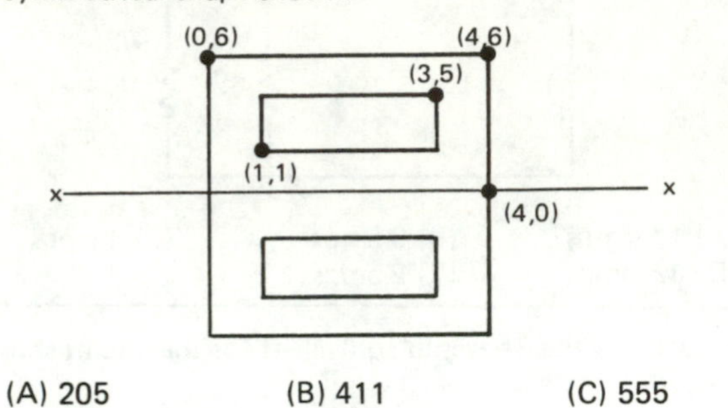

(A) 205 (B) 411 (C) 555
(D) 473 (E) 288

75. What is the equilibrium angle θ if the one pound rod length is 24"? Assume all surfaces are smooth.

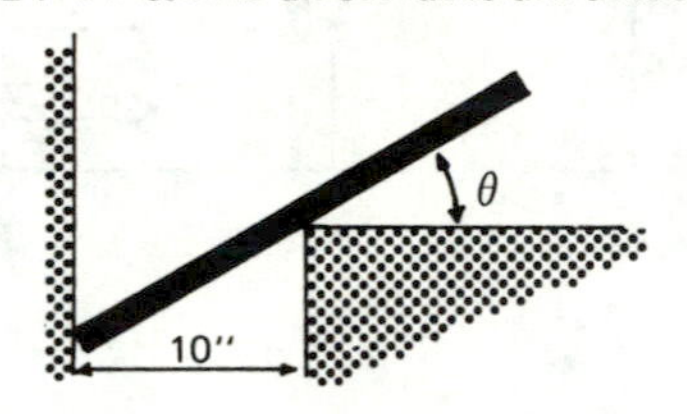

(A) 19.8° (B) 65.4° (C) 33.6°
(D) 56.4° (E) 39.8°

76. What is the force in member AC in terms of the load P?

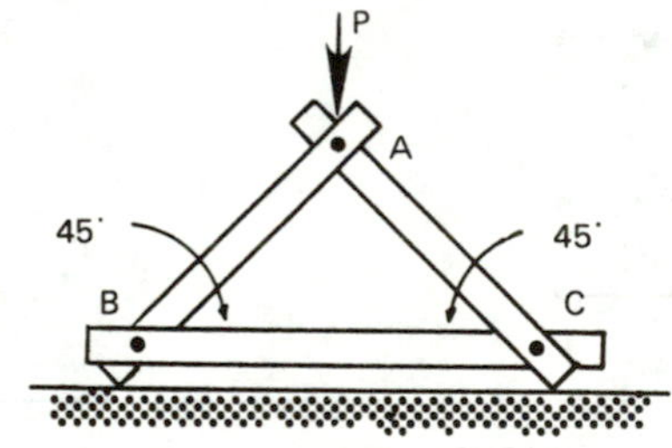

(A) .866P (B) .71P (C) P
(D) .5P (E) cannot be found

77. What force F is required to lift the 50 pound load? Consider all strands to be parallel.

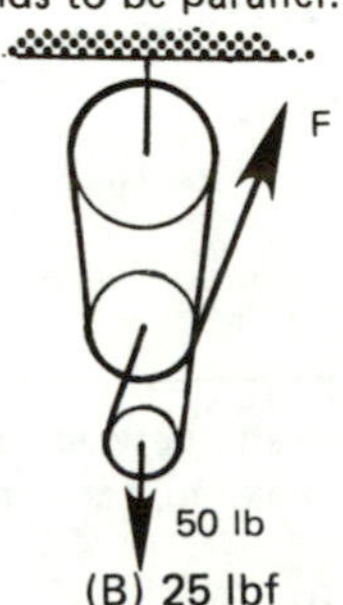

(A) 50 lbf (B) 25 lbf (C) 16.7 lbf
(D) 12.5 lbf (E) 8.3 lbf

78. Find the stress in the member marked by an 'X'.

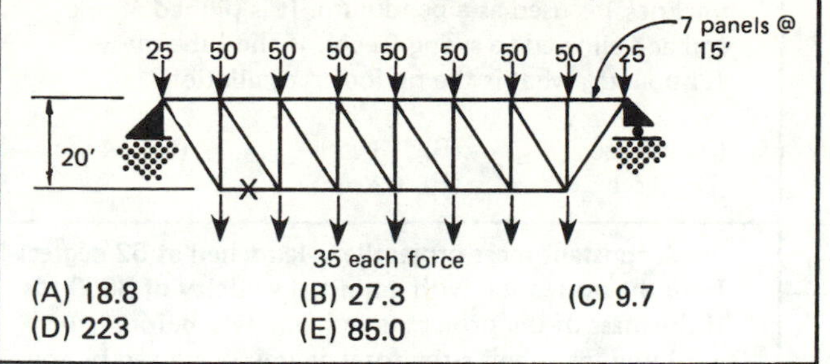

(A) 18.8 (B) 27.3 (C) 9.7
(D) 223 (E) 85.0

79. What is the force in member AB?

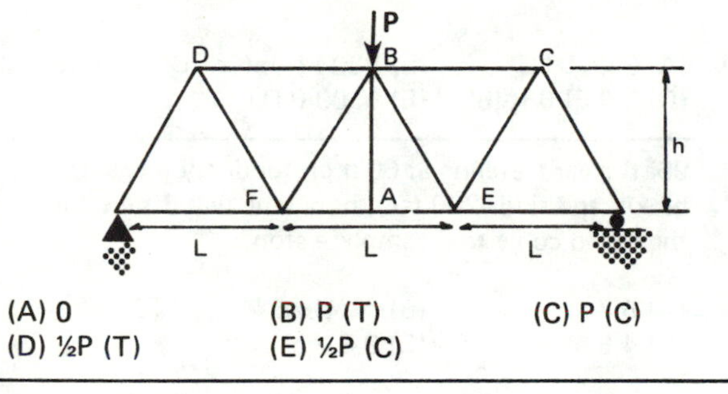

(A) 0 (B) P (T) (C) P (C)
(D) ½P (T) (E) ½P (C)

80. Referring to the truss in problem number 79, what is the force in member BC?

(A) 0 (B) LP/H (C) (C) LP/H (T)
(D) PL/2H (C) (E) PL/2H (T)

81. Neglecting the weight of the beam itself, what is the maximum moment acting on a cantilever beam of length 20 feet. if a concentrated load of 100 pounds acts 4 feet from the free end?

(A) 400 ft-lbf (B) 200 ft-lbf (C) 160 ft-lbf
(D) 2000 ft-lbf (E) 1600 ft-lbf

82. The approximate vertical force component in member BC is

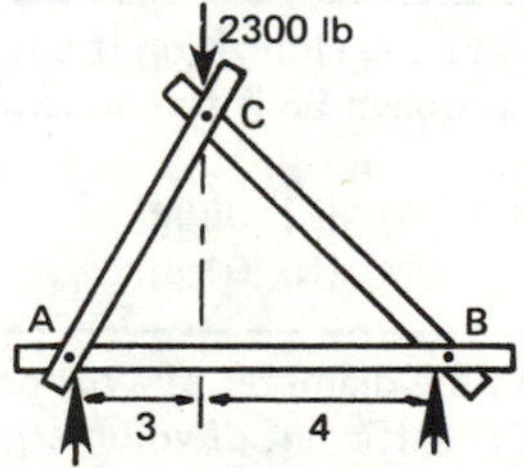

(A) 3600 lb (B) 2300 lb (C) 1300 lb
(D) 1000 lb (E) indeterminate

83. What is the frictional force between the block and the ramp? The coefficient of static friction is .2. The coefficient of dynamic friction is .15.

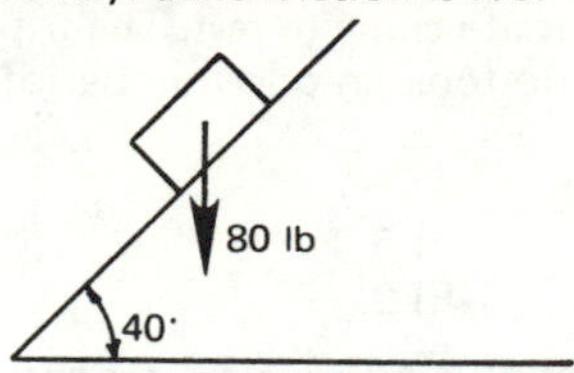

(A) 16 lbf (B) 12 lbf (C) 9 lbf
(D) 6 lbf (E) 8 lbf

84. What is the reaction at A?

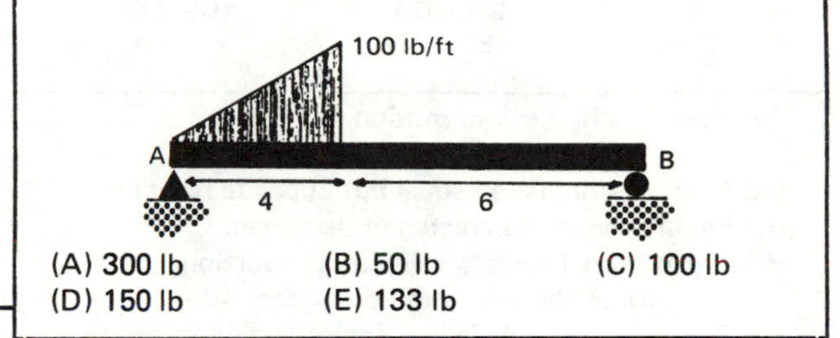

(A) 300 lb (B) 50 lb (C) 100 lb
(D) 150 lb (E) 133 lb

DYNAMICS

85. Given the component velocities below, what is the resultant velocity at t = 4?

$$v_x = t^3 - t$$
$$v_y = 3 - t^2$$

(A) 60 (B) 13 (C) 27
(D) 61 (E) 73

86. How far will an object drop in 13 seconds, starting from rest and neglecting air friction?

(A) 419 ft (B) 5442 ft (C) 2721 ft
(D) 5280 ft (E) cannot be found

87. What is the x component of velocity after impact if the coefficient of restitution is .8?

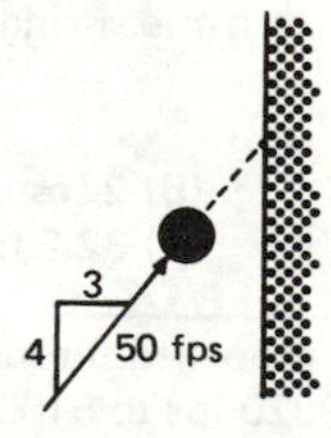

(A) 30 fps (B) 40 fps (C) 50 fps
(D) 24 fps (E) 32 fps

88. A car accelerates uniformly from rest to 8K mph in K/5 minutes. K is a fixed constant. The car continues at that speed for K minutes, then decelerates uniformly taking K/5 minutes to come to rest. The trip length was K-1 miles. The trip took an odd number of minutes. What is K?

(A) 1.25 (B) 5 (C) 4
(D) 3 (E) 2

89. The velocity of a particle at time t is

$$v(t) = 12t^4 + (7/t)$$

What distance is traveled between t = .2 and t = .3?

(A) +98.1 (B) -98.1 (C) -2.84
(D) +2.84 (E) +8.42

90. Newton's first law of motion says

(A) Every action has an equal but opposite reaction.
(B) Energy cannot be created or destroyed.
(C) Gravitational force is inversely proportional to the square of the separation distance.
(D) Particles with unbalanced forces remain in a state of rest.
(E) The momentum of a particle remains unchanged in the absence of unbalanced forces.

91. The value of escape velocity is determined by

(A) multiplying the object mass times the reciprocal of the planet's gravitational constant.
(B) integrating Newton's law of universal gravitation from the planet's surface to an infinite distance.
(C) experimentation and wind tunnel testing.
(D) assuming an infinite mass and negligible separation between bodies.
(E) assuming zero mass and finite separation between bodies.

92. A spring has a spring constant of 10 lbf/in. It is compressed 5 inches. The spring is released and pushes against a free projectile with a mass of one pound. What is the projectile velocity immediately after losing contact with the spring?

(A) 11.6 fps (B) 2 fps (C) 64.4 fps
(D) 25.9 fps (E) 32.2 fps

93. A rocket is propelled through a vacuum. It changes its velocity from 9020 fps to 5100 fps in 48 seconds. What power is required to accomplish this if the rocket's mass is 13,000 slugs?

(A) 5.3 EE5 BTU/sec (B) 965 hp (C) 1.36 EE7 hp
(D) 2.08 EE9 hp (E) 3.78 EE6 hp

94. A hollow tube (1 foot long, ½" O.D., .125" wall thickness) is used as a pendulum. It is pinned at one end and allowed to swing freely. If the tube mass is 1.1 pounds, what is the period of oscillation?

(A) .45 sec (B) .13 sec (C) .14 sec
(D) .9 sec (E) .02 sec

95. A constant mass projectile is launched at 52 degrees from the horizontal with an initial velocity of 3600 fps. If the mass of the projectile immediately before launch is 32 pounds, what is the total energy possessed by the projectile at t = 13 seconds? Neglect all forms of friction.

(A) 64.4 BTU (B) 8300 ft-lbf (C) 1.4 EE6 ft-lbf
(D) 5.0 EE6 ft-lbf (E) 8300 BTU

96. If a car traveling at 60 mph suddenly locks its brakes and skids 200 feet, how long will it take for the car to come to a complete stop?

(A) 6.7 sec (B) 44.4 sec (C) 3.3 sec
(D) 4.5 sec (E) 9.8 sec

97. What momentum does a 40 lbm projectile possess if it is moving at 420 mph?

(A) 16,800 lbf-sec (B) 24,640 lbf-sec (C) 770 lbf-sec
(D) 520 lbf-sec (E) 793,400 lbf-sec

MECHANICS OF MATERIALS

98. A 30" long rod (E = 3 EE7 psi, α = 6 EE-6 in/in-°F) with a 2 square inch cross section is fixed at both ends. If the rod is heated to 60°F above the neutral temperature, what is the stress?

(A) 36,000 psi (B) 109 psi (C) 10,800 psi
(D) 1.8 EE9 psi (E) 57,070 psi

99. What will be the elongation if one end of the rod described in problem 98 is free to expand?

(A) 3.6 EE-4 in (B) .08 in (C) 5.4 EE-4 in
(D) .03 in (E) .01 in

100. At 70°F, the diameter of a balloon is 10 inches. The balloon's coefficient of volumetric expansion is 2 EE-3 1/°F. What is the balloon's temperature when its diameter is 12 inches?

(A) 93°F (B) 77°F (C) 434°F
(D) 115°F (E) 364°F

101. Vickers, Knoop, and Brinell are all names of

(A) Nobel prize winners in metallurgy.
(B) magnetic constants.
(C) thermodynamic constants.
(D) hardness tests.
(E) Chi-squared statistics.

102. What is the elongation in the cable if F = 1000 pounds? The cable's cross sectional area is 2 square inches. Its modulus of elasticity is 1.5 EE6 psi.

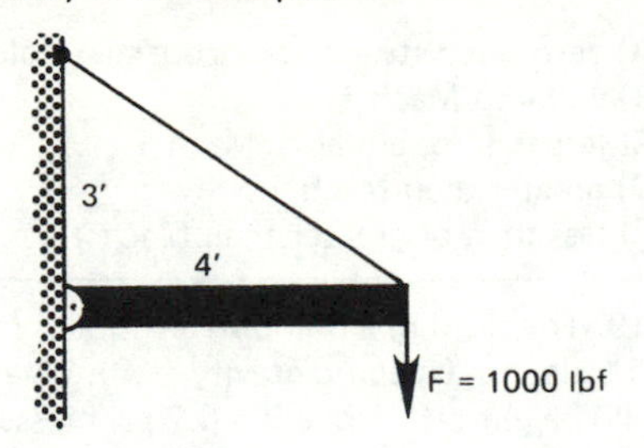

(A) .0028 in. (B) .03 in. (C) .02 in.
(D) .04 in. (E) .57 in.

103. A simply supported beam carries a single concentrated load at its center. If its slenderness ratio is 150, it is most likely to fail

(A) where shear is zero.
(B) where the moment is zero.
(C) where the shear is maximum.
(D) at a support.
(E) where the slope of the deflection curve is zero.

104. For a fixed curing time, the ultimate strength of concrete

(A) increases with a decrease in water content.
(B) decreases with a decrease in water content.
(C) is independent of water content if cured for a sufficiently long time.
(D) is independent of curing pressure.
(E) none of the above.

105. A 30 foot long steel column, 2 inches x 2 inches, is one-third embedded in concrete. If the upper end is free to move and a safety factor of 4 is used, what is the maximum permissible load?

(A) 1713 lb (B) 107 lb (C) 190 lb
(D) 762 lb (E) 428 lb

106. A rectangular beam 4" wide and 6" high is subjected to a shear of 7000 pounds at a particular location. The beam is constructed of 1100 aluminum. What is the maximum shear stress at that particular location?

(A) 290 psi (B) 440 psi (C) 520 psi
(D) 660 psi (E) none of the above

107. A 1020 carbon steel rod is ¼" in diameter and 6" long. What torque must be applied to twist the rod 8 degrees?

(A) 116 in-lbf (B) 116 ft-lbf (C) 41,760 in-lbf
(D) 3480 ft-lbf (E) 268 in-lbf

108. A round steel bar with a 9 square inch cross section is rigidly clamped at each end of its 9 inch length. A 50,000 pound load is applied 3 inches from its left end. What is the reaction at the left end?

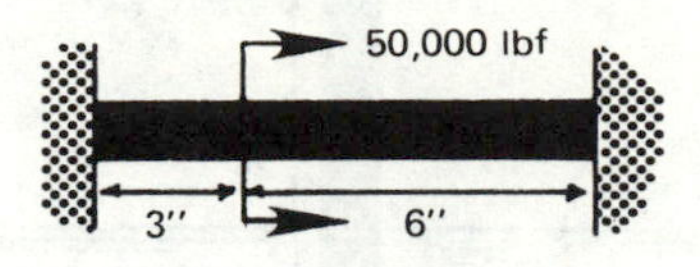

(A) 17,000 lbf (B) 33,000 lbf (C) 25,000 lbf
(D) 19,000 lbf (E) 50,000 lbf

109. Which of the following is not a method by which a single lap rivit joint usually fails?

(A) The rivits shear.
(B) The rivits fail in bearing.
(C) The plate fails in bearing.
(D) The plate fails in tension
(E) The rivits fail in tension.

110. What is the maximum flexural stress at a distance x from the free end of a cantilever beam supporting a tip load P?

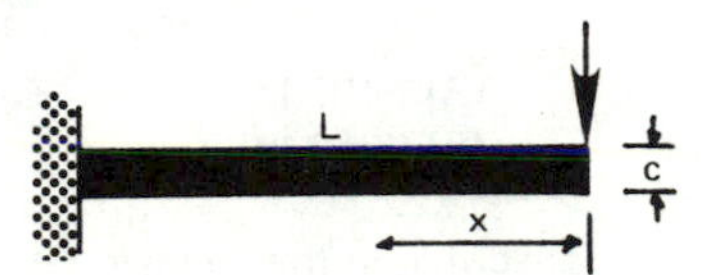

(A) Pxc/I (B) Pxc/2I (C) Pc/2I
(D) PcL/2I (E) Pxc/2EI

FLUID MECHANICS

111. A venturi is used to measure air velocity. A 1/5 scale model is built and water is used as the test fluid. Viscosity of the air is 3.8 EE-7 lb-sec/ft^2. Viscosity of the water is 2.05 EE-5 lb-sec/ft^2. What will be the ratio of the model to actual velocities observed?

(A) .3 (B) 5 (C) 25
(D) 270 (E) 11

112. A 500 foot long surface vessel is modeled at 1:50. What speed must the model travel if a 25 mph similarity is desired?

(A) 7.07 fps (B) 3.54 fps (C) 5.2 fps
(D) 7.3 fps (E) 5.0 fps

113. What is the hydraulic radius of an equilateral triangle (vertex down) open channel flowing at half capacity and with maximum depth of 3 feet?

(A) .6 (B) .65 (C) .7
(D) .75 (E) .8

114. A 24 inch long rod floats vertically in water as shown. It has a 1 square inch cross section and a specific gravity of 0.6. What length (L) is submerged?

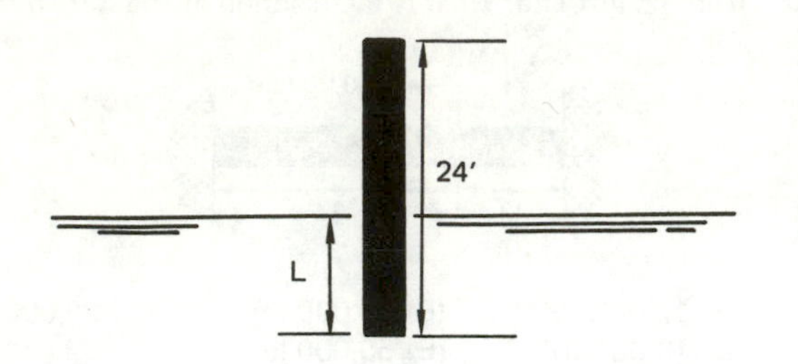

(A) 9.6"
(B) 14.4"
(C) 18.7"
(D) 18.0"
(E) 2.4"

115. A tank is filled with water to a depth of 10 feet. What is the total force on the gate shown below?

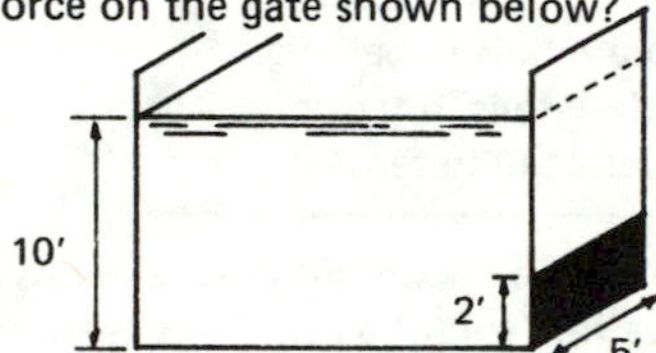

(A) 560 lbf
(B) 5400 lbf
(C) 5600 lbf
(D) 6000 lbf
(E) 6240 lbf

116. What force is acting on the inclined section of the water tank below? Express your answer in pounds per foot width of tank.

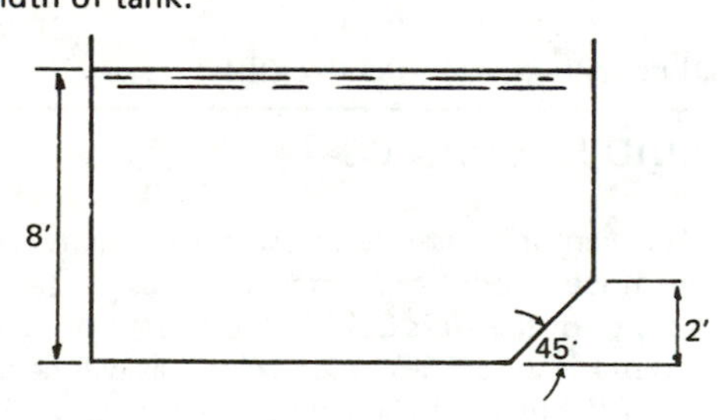

(A) 1740 lbf/ft
(B) 870 lbf/ft
(C) 1240 lbf/ft
(D) 1590 lbf/ft
(E) 2280 lbf/ft

117. When a thin-bore, hollow glass tube is inserted into a container of mercury, the surface of the mercury in the tube

(A) is level with the surface of the mercury in the container.
(B) is below the container surface due to cohesion.
(C) is below the container surface due to adhesion.
(D) is above the container surface due to cohesion.
(E) is above the container surface due to adhesion.

118. The velocity of a fluid in the throat of a properly designed convergent/divergent nozzle is

(A) zero since stagnation properties hold.
(B) equal to Mach 1.
(C) less than or equal to Mach 1.
(D) greater than Mach 1.
(E) less than or greater than Mach 1.

119. The fluid in a manometer tube is 60% water and 40% alcohol (specific gravity = .8). What is the manometer fluid height difference if a 6.2 psi pressure difference is applied across the two ends of the manometer?

(A) .92 in.
(B) 361 in.
(C) 187 in.
(D) 215 in.
(E) 172 in.

120. A semi-circular open channel is

(A) inferior to a flat-bottom rectangular channel of equal flow area in terms of flow quantity.
(B) superior to a V-trough in terms of cost per unit length.
(C) inferior to a flat-bottom, trapezoidal channel of equal flow area in terms of flow quantity.
(D) inferior to a V-trough channel of equal flow area in terms of flow quantity.
(E) inferior to the V-trough, rectangular, and trapezoidal channels in terms of cost per unit length.

121. What is the expected head loss per mile of closed circular pipe (17" inside diameter, friction factor = .03) when 3300 gpm of water flow under pressure?

(A) 38 ft
(B) .007 ft
(C) 1.9 EE7 ft
(D) 3580 ft
(E) 40 psf

122. A perfect venturi with throat diameter of 1.8 inch is placed in a pipe of 5 inch inside diameter. 80 pounds of water flow through the pipe each second. What is the difference between the pipe and venturi throat static pressures?

(A) insufficient data
(B) 34.8 psi
(C) 4980 psi
(D) 29.9 mm Hg
(E) 72.32 fps

123. The Froude number is a

(A) ratio of two irrational prime numbers.
(B) ratio proportional to the coefficient of heat transfer divided by the thermal conductivity.
(C) ratio proportional to the specific heat at constant pressure divided by the thermal conductivity.
(D) ratio proportional to the velocity divided by the square root of the characteristic dimension.
(E) ratio proportional to the mass flow rate per unit area divided by the viscosity.

124. Turbulent flow is said to exist when

(A) pipe friction exists.
(B) streamlines are parallel to the direction of flow.
(C) streamlines are perpendicular to the direction of flow.
(D) the velocity distribution in a cross section is essentially uniform.
(E) cross-current, pulsatory eddies are absent.

COMPUTER SCIENCE

125. A bootstrap loader is

(A) an early card reader.
(B) a form of compiler.
(C) a short sequence of instructions which loads a larger program.
(D) a temporary storage register for loading machine language instructions.
(E) a read-only memory containing standard algorithms.

126. A Z80-based microprocessor has 15 storage registers of 16 bytes each allocated to a particular function. How many bits are contained in the registers?

(A) 15 (B) 16 (C) 240
(D) 128 (E) 1920

127. Which of the following types of logic is the fastest?

(A) CMOS (B) RTL (C) TTL
(D) discrete components (E) MSI

128. The FORTRAN variable **STOCK** defined below will most likely

```
INTEGER STOCK*2
STOCK = 14
STOCK = STOCK*3000
```

(A) be equal to 42000.
(B) cause underflow.
(C) cause an interrupt.
(D) be set equal to 9232.
(E) be set equal to 9233.

129. What is the value of the FORTRAN variable **IBJEKT**?

```
MUCH = 7+9*6/9/4**4
IBJEKT = 2*MUCH
```

(A) 50.57 (B) 50 (C) 28561
(D) 14.02 (E) 14

130. What is the binary number 1101001101 in base 16?

(A) 34D (B) 3413 (C) 3415
(D) 845 (E) 1515

131. What is the one's complement of the binary number 10110?

(A) 10111 (B) 10101 (C) 01001
(D) 01010 (E) 01000

132. Which of the following is not an advantage of a subroutine in a FORTRAN program?

(A) More initial coding is usually required.
(B) The subroutine may be called from different parts of the main program with different arguments.
(C) The subroutine may be written independently of the main program.
(D) The subroutine may be written in machine language to speed execution.
(E) Once written, a subroutine may be used with any calling program.

SYSTEMS ENGINEERING

133. *Systems Engineering* considers or involves which of the following?

1. goals
2. alternatives
3. models
4. computers
5. sensitivity analysis

(A) 1, 2, 3, 4, 5
(B) 1, 2, 3, 5
(C) 1, 2, 3, 4
(D) 1, 2, 4, 5
(E) 3, 4, 5

134. A transfer function F(s) is given below. A pole of F(s) is most nearly

$$F(s) = \frac{s^2 - 2s + 5}{s(s+2)}$$

(A) +2 (B) -2 (C) +3
(D) $1 + i2$ (E) $1 - i2$

135. The Laplace transform of a step function of height h is

(A) 1/(s+h) (B) 1/s (C) 1/h
(D) h/s (E) h

136. What is the critical path through the network shown below?

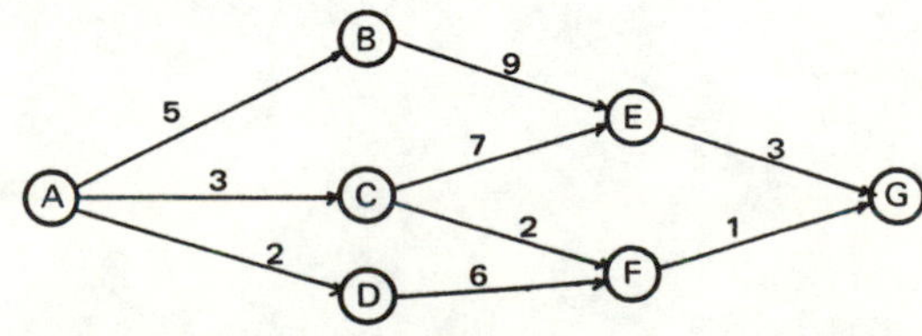

(A) ACEG (B) ABEG (C) ACFG
(D) ADFG (E) ABECFG

137. What is the overall gain (x_o/x_i) of the positive feedback system shown below?

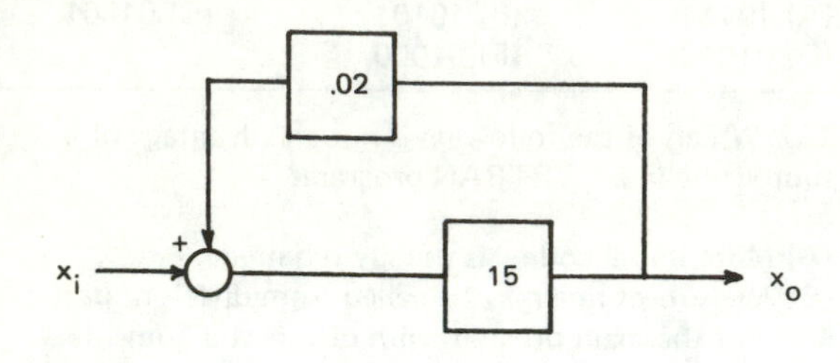

(A) 750.0 (B) 14.98 (C) 15.02
(D) 11.54 (E) 21.42

138. What is the sensitivity of the system shown in problem 137?

(A) 143% (B) 77% (C) 3%
(D) 18% (E) 47%

139. In reliability theory, the hazard rate for a negative exponential distribution

(A) increases with time.
(B) decreases with time.
(C) is independent of time.
(D) is an imaginary function of time.
(E) none of the above

140. What is **R** in the FORTRAN statement below?

```
R = AMOD(27.642, 12)
```

(A) 2 (B) 2.3035 (C) 331.704
(D) .704 (E) .3035

STOP!

You have completed the morning session of the E-I-T exam. If you have time remaining, you should check your work. You may not go on to the afternoon session of the exam until you are told to do so by your proctor.

If you wish, you may turn in your answer sheet and leave the examination room. However, you may not take this examination booklet with you.

Please be considerate of other examinees.

STOP! Do not turn this page. If you have time left over, check your work. Check to be sure that all of your responses on the answer sheet are dark and completely fill the bubbles.

Do not look at the answers until you have finished taking this exam. When you have finished taking this exam, or when you have used up all of your allotted time, turn the page and locate the answers to the problems in this exam.

After you have compared your answers to the answer key, determine your score according to the grading rules listed on page 1. If you have scored poorly on the exam, you may want to consider additional study before taking your actual E-I-T exam.

Do not look at the answer key until you are ready to score your own examination responses.

E-I-T FINAL EXAM ANSWER KEY

1. e
2. d
3. b
4. d
5. e
6. e
7. a
8. d
9. d
10. a
11. b
12. a
13. d
14. e
15. a
16. b
17. a
18. c
19. a
20. e
21. e
22. e
23. d
24. c
25. a
26. c
27. a
28. d
29. c
30. e
31. d
32. c
33. e
34. b
35. b
36. e
37. c
38. c
39. c
40. a
41. a
42. b
43. e
44. d
45. b
46. b
47. e
48. a
49. a
50. c
51. c
52. e
53. b
54. d
55. c
56. d
57. b
58. d
59. b
60. b
61. e
62. a
63. c
64. b
65. c
66. e
67. d
68. c
69. c
70. b
71. a
72. c
73. a
74. b
75. a
76. b
77. b
78. d
79. a
80. d
81. e
82. d
83. c
84. d
85. d
86. c
87. d
88. b
89. d
90. e
91. b
92. d
93. c
94. d
95. e
96. d
97. c
98. c
99. e
100. c
101. d
102. b
103. e
104. a
105. e
106. b
107. a
108. b
109. e
110. b
111. a
112. c
113. d
114. b
115. c
116. c
117. b
118. c
119. c
120. e
121. a
122. b
123. d
124. d
125. c
126. e
127. c
128. c
129. e
130. a
131. c
132. a
133. a
134. b
135. d
136. b
137. e
138. a
139. c
140. e

Indispensable Exposure to the

ENGINEER-IN-TRAINING EXAMINATION

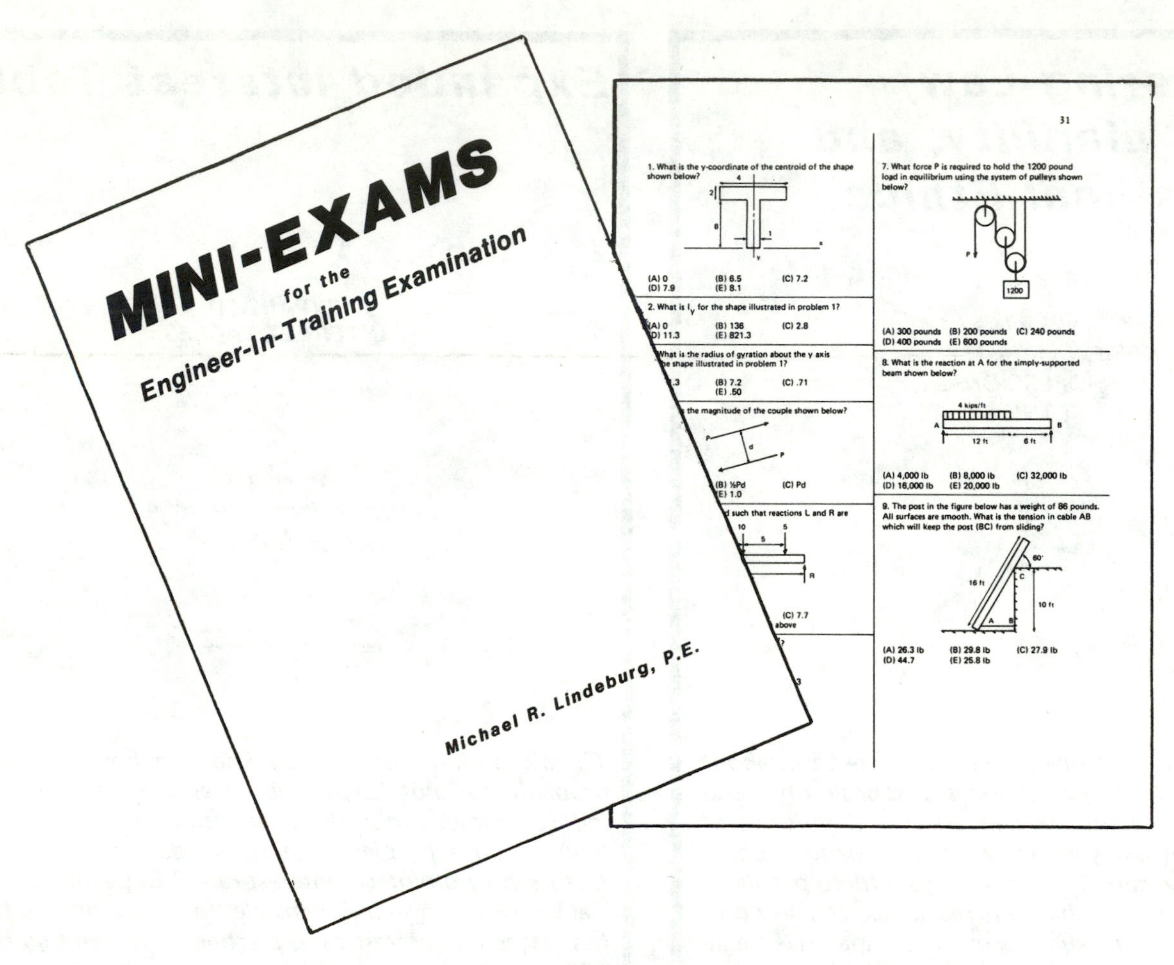

Here is what you need to overcome the most difficult aspect of the E-I-T examination: speed. Only repeated dry-run exposure to the examination format can condition you to the pace required to get through all problems in 8 hours.

The *MINI–EXAMS* consists of 17 different 30-minute examinations, each on a different subject. The exam format, instructions, and answer sheet rigidly parallel the actual exam. Solutions to all problems are provided in the back of the book.

When you go to take the actual exam, you will have seen its complexity 17 times before. You will know the pace you must maintain. And, best of all, you will have developed the ability to recognize and skip problems that will waste your time.

8½" X 11", soft cover. 82 pages, printed front and back. *$15.30* (includes postage), plus 6½% for California residents. (ISBN 0–932276–34–2)

Inquire at your local technical or university bookstore. Or order directly.

PROFESSIONAL PUBLICATIONS, INC.
Department 77
1250 Fifth Avenue
Belmont, CA 94002
(415) 593-9119